FROM THE BIG BANG TO CONSCIOUSNESS

Life's Challenge to Entropy

Chiaretto Calò

From the Big Bang to Consciousness

Copyright © 2024 by Chiaretto Calò

All rights reserved.

First edition September 2024

Cover image licensed from Shutterstock.

English Paperback Edition

ISBN 9798337951751

No part of this publication may be reproduced, distributed, or transmitted in any form or by any means, including photocopying, recording, or other electronic or mechanical methods, without the author's prior written permission, except as permitted by copyright law.

For permission requests or to report errors, contact the author at chiarettocalo@gmail.com

*To my ancestors,
that with immense vital effort
have made it possible
my arrival here,
to find out
and narrate the universe.*

CONTENTS

pg.

Introduction   13

PART ONE

THE FOUNDATIONS.

Chapter 1: The laws of the Universe and the rebellion of life

1	The Laws of the Universe	31
2	Understanding thermodynamics	34
3	The Second Law and the mystery of the origin of the universe	37
4	Life: an anomaly?	42
5	The concept of dissipative structures	44
6	Life as a dissipative structure	46
7	The role of energy flow	48
8	Evolution: a response to entropy	49

Chapter 2: The origin of life

1	The primordial soup	63
2	The role of energy in the origin of life	67
3	The emergence of order	69
4	The first dissipative structures	71
5	From the simple to the complex: the evolutionary leap	73
6	The world of RNA	76

7	The advent of DNA	*77*
8	The birth of metabolism	*78*
9	The first cells	*80*

Chapter 3: The language of life: DNA

1	The structure of DNA	*91*
2	DNA: thermodynamic miracle?	*92*
3	The role of DNA in energy dissipation	*94*
4	The evolution of the genetic code	*95*
5	Genetic variation	*96*
6	The role of sexual reproduction	*98*
7	Epigenetics: another level of complexity	*99*
8	DNA damage and repair	*101*
9	The human genome	*102*
10	The future of DNA: genetic engineering	*104*

PART TWO

THE TRANSFORMATION

Chapter 4: The Universe's strategy for equilibrium

1	The mechanisms of evolution	*117*
2	Survival of the fittest	*119*
3	Evolution as energy optimization	*122*

4	The role of mutation	124
5	The rise of complexity	127
6	The Cambrian explosion	129
7	Evolutionary arms races and energy dissipation	131
8	The emergence of intelligence	132
9	Human evolution	135
10	The future of evolution	138

Chapter 5: Memory: a thermodynamic framework of experience

1	The nature of memory	149
2	Memory and brain structure	151
3	The role of energy in memory formation	152
4	Memory, learning and entropy	154
5	The evolution of memory systems	155
6	Memory and consciousness	157
7	The thermodynamics of forgetting	159
8	Memory disorders: thermodynamic perspective	160
9	The future of memory: improvements and implications	161
10	Memory and immortality	163

Chapter 6: The human mind: a thermodynamic powerhouse.

1	The complexity of the human brain	175
2	Brain energy consumption	176
3	The brain's role in energy dissipation	177
4	Consciousness: a thermodynamic conundrum	179
5	The evolution of human intelligence	180
6	Psychology of emotions	182

7	The thermodynamics of decision making	*183*
8	Mental disorders: a thermodynamic view	*184*
9	The future of the human mind: AI and beyond	*186*
10	The thermodynamics of the singularity	*187*

PART THREE

THE ASCENSION

Chapter 7: Societies and cultures: the thermodynamics of human interaction

1	The emergence of humane societies	*205*
2	The role of energy in social structures	*206*
3	The thermodynamics of culture	*208*
4	Language and entropy	*210*
5	The evolution of societies	*211*
6	The rise and fall of civilizations	*215*
7	War and Peace	*217*
8	Globalization	*219*
9	The thermodynamics of the Anthropocene	*222*
10	The future of societies	*224*

Chapter 8: Technology: humanity's thermodynamic tool

| 1 | The dawn of technology | *233* |

2	The role of energy in technological progress	*234*
3	The thermodynamics of innovation	*236*
4	Technology and energy dissipation	*239*
5	The information age: a thermodynamic revolution	*241*
6	The thermodynamics of the Internet	*244*
7	The energy cost of technology	*246*
8	The future of technology	*249*
9	Space exploration	*253*
10	Technology and the Second Law: a final frontier	*256*

Chapter 9: The future: Life, the universe and entropy

1	The distant future	*265*
2	The heat death of the Universe	*267*
3	Life's struggle against the inevitable	*269*
4	The future of life: a thermodynamic forecast	*271*
5	The role of intelligence in the future of the universe	*274*
6	The possibility of life elsewhere in the universe	*276*
7	The thermodynamics of alien life	*279*
8	The future of humanity	*282*
9	Immortality	*284*
10	The final word	*287*

PART FOUR

THE INVISIBLE

Chapter 10: Hidden dimensions

1	The vibration of wave-particles	*301*
2	Space-time paradoxes	*302*
3	Holographic Universe	*303*
4	The law of syntropy	*308*
5	Quantum vacuum structure	*309*
6	Syntropic force fields	*311*
7	Collective consciousness	*312*
8	Toward a holistic understanding of the Universe	*313*

Conclusion	*321*
Index of main authors	*327*
Index of main concepts	*333*

From the Big Bang to Consciousness

INTRODUCTION

Why do we remember the past but not the future? Why does life emerge as a *current of order* within a tide of apparent disorder? Why does human existence follow a path that, from birth, from the existent, inexorably heads toward an end, a *non-existent*, an ultimate direction that seems to contradict the fullness of essence?

In the *Timaeus*, Plato ventures into one of the earliest philosophical explorations of the cosmos and life by placing human beings and their temporal experience in a broader and deeply interconnected cosmic context. From the Platonic perspective, time is not a mere succession of events but is a *moving image* (εἰκών κινούμενος) of eternity (αἰώνιος) dwelling in the celestial sphere. The *eternal heaven* (τὸν αἰώνιον οὐρανόν) is the perfect and unchanging pattern, while time – *Chronos* (χρόνος) – is its imperfect and mutable copy flowing in the sensible realm (Timaeus, 37d-38a). Life is thus a temporal reflection of absolute eternity and struggles tirelessly against entropic decay. In the following passage (Timaeus, 39e), the philosopher describes the *divine providence* (πρόνοια θεῖα) that *orders* the universe and *brings disorder into order*, a principle that can be interpreted as the opposition to entropic chaos. Life, imbued with an *ordering principle*, thus stands against the current of entropy, almost as if it had been divinely commissioned to maintain an island of order within the sea of chaos. The philosopher tells us how the *immortal soul* (ἀθάνατος ψυχή) is responsible for introducing order from the eternal to the temporal order (Timaeus, 35a-b). If life can be seen as a force that opposes entropy, then perhaps there is a deeper message hidden in the laws that govern the cosmos, a message that Plato seeks to unveil through myth and *logos*. Perhaps, as he suggests, there is a *demiurge* (δημιουργός) who shapes the cosmos with purpose and measure (Timaeus, 28a), indicating how the laws of the universe are marked by a rationality that transcends mere physicality.

Time, an elusive entity that flows inexorably from what has been to what will be, is guided by the arrow of entropy, that thermodynamic principle that paints a picture of inevitable and growing disorder. Yet, at the heart of this unidirectional flow, life emerges as a cunning force, a vortex of order and complexity that defies the dominance of *Chronos*. In a sense, life disguises itself as an exception to the universal laws of the gods, a deception that allows ordered complexity to flourish amid chaos. In the meditation on time, the concept of memory is projected toward the past, not toward the future, because only what has been stands out in being, while the future lies in the shadows of *non-being*. Time, in its essence, is not a sequence of instants but is *being itself manifested* through entropic phenomenon.

Entropy must be understood not as a disorder but as the only tangible manifestation of time itself, of the Being that is appearing and disappearing in the flow of epochs. Human existence, a journey from beingness to nonbeing, is then revealed as a parable that reflects the truth of time: *we are beings in time*, and our journey from beingness to nonbeing is a manifestation of being itself. The non-essence toward which we head is not a negation but a confirmation that the *being is revealed in disappearance, just as in appearance.* Time does not deny essence; on the contrary, it realizes it. The realization-in-act of power is the phenomenological foundation on which entropy rests, the *visible face of time* which, as science indicates, measures the most probable *irreversibility of becoming*, the direction in which the cosmos, in its totality, moves. Entropy and time are thus intertwined: one is the phenomenal expression of the other, a becoming that in itself has no direction except that imprinted by being that becomes time. Being, eternal and unchanging in its concealment, in its withdrawal, becomes time, history, life, and death in the becoming of the forms that are its epiphanies.

Entropy is a measure of uncertainty and ignorance. Yet, life preserves and transcends information: a tree encodes in its DNA the memory of a distant past; a child learns a language and culture of generations past; a brain adapts, learns, and prevents, defying the eternal flow of entropy. In recognizing this, the question about memory and the future is transformed: it is not a memory that extends only backward and not in an inaccessible future. Our being-in-time is a permanent becoming toward what we are: beings destined to manifest through time, entropy, and life simultaneously. Memory binds with being that becomes time, the future with entropy that reveals the truth of being: that *nothing is lost, nothing is annihilated, but everything returns in being that unfolds and retracts in the eternal play of the cosmos*.

Insofar as it is unknown, the future seems to conceal within itself the apparent otherness of the ontological foundation in a boundless extension, an infinity that eludes the nets of finiteness that advance in time but resists our knowing. The being behind the appearing is revealed in the horizon of temporality and human memory, extending backward in time and not in the future, bearing witness to our *original fall* into its oblivion. An oblivion that prevents us from anticipating the future to the same extent that we remember the past. And human existence, in the journey from being to death, only follows the course of the truth of our time, which sees becoming as the only real and nothingness as the ultimate destiny of everything. But the truth is otherwise: being evades becoming, and death is but the appearance that the foundation offers of itself in becoming. Our experience, which relies on Cartesian coordinates and is limited to the empirical, can only approach the infinite without ever being able to embrace it fully. The universe, in its proceeding, simultaneously reveals and conceals its goal, which is not a place in the future but the constant presence of being retreating into the past. The drama of man, immersed in the cosmos yet eternally distant from penetrating its arcane, is thus the *drama of the oblivion of being*.

Contemporary physics, with its equations and models, tries to penetrate the arcane, but every answer it seems to offer is actually an invitation to recognize the insufficiency of every answer and the beginning of a new question. We are called to acknowledge the universe's existence but also to listen to a tragic *silence of being* that lies behind every phenomenon, law, and equation - for in that silence lies the truth that precedes and surpasses all our attempts at formulation. The chain of questions and partial answers is but the wandering path of the human who, forgetful of being, loses himself in the endless game of appearances. The human condition, suspended between the certainty of ignorance and the hope of knowledge, is a continuous struggle against the boundaries of the intellect. Our memory extends into the past, an archive of events that we have the power to recall, but the future remains a blank sheet that only the act of existing can transcribe. Death, moreover, is but the seal of the existential condition, the end of our personal contribution to cosmic becoming, a legacy to future generations with the task of continuing the narrative.

Our ancestors immersed themselves and were at one with the starry sky from which they read the fate and will of the gods. Now, although the enchantment of mystery persists, the sacred has been replaced by science and technology, although the ongoing search for truth is not to be considered any less sacred. In an age of proclaimed *post-metaphysics*, we find ourselves even more alienated from understanding the ultimate purposes of the universe. In the *age of technology*, where the human artifact seems to surpass the human creator in capacity and complexity, the question about the universe's movement becomes charged with additional meanings. Science becomes as much a pathway into mystery as myth was; a new myth, however, that does not quench but exacerbates our awe, making us more and more aware of the vastness of our ignorance the deeper we deepen our understanding.

What relationship exists between our awe and the nature of mystery? It is a question that accompanies us as we continue to peer into the darkness of the universe, trying to understand our place in it, discern the nature of our collective journey into the unknown, and make sense of our transit through the enchantment of the cosmos. The search for the universal, then, manifests itself in the tension between entropy, that tendency toward increasing disorder described by the law of thermodynamics, and life, which in an almost heroic way pursues order, structure, and complexity.

Entropy, the symbol of an inescapable nature that erodes forms and disrupts syntheses, stands out against creative order, a *logos that infuses meaning into chaos*. In physics, the principle of order could be paralleled by the force that imposes coherence on the universe.

And so, in our meditation, we find that *the universe itself is constituted as a challenge of self-knowledge*, a hymn to the *inherent creativity of reality*, inviting deeper reflection on the meaning of existence and the laws that govern the cosmos. Are we ready to embark on the journey of understanding? It is a mystery that calls us to contemplate not only how life can exist in a universe governed by thermodynamics but also what role it plays in the larger order of existence, an order that transcends physical laws and plunges into the depths of the mystery of being. One of the most intriguing aspects is the contrast between entropy - the tendency toward disorder described by the second law of thermodynamics - and life - which constantly seeks order and complexity. Entropy, often defined as *the measure of disorder in a physical system*, is one of the constants of nature. It dances in chaos, breaking down structures and generating disorder. But as in any great dance, there is no one dancer. Quantum physics teaches us that reality is far more complex than we can perceive, and there may be an *invisible partner* of entropy that we do not yet fully understand. Who, then, could be its invisible partner? Next to entropy, is there perhaps a principle of order, of a creative kind, a *logos* that *in-forms* and gives meaning to chaos? In the world of physics, the principle of order can be compared to the fundamental force that governs the universe. Can the *logos*, therefore, constitute the path to understanding?

Philosophical reflection on the concept of life is deeply intertwined with the roots of ancient thought. Plato spoke of the human soul as a *divine spark* intrinsic to the human being. The Greeks, with their inquiry into *bios*, not only referred to life in the biological sense but also to the quality and value of existence. Resting on such foundations, philosophical reflection on *bios* reveals itself as a simple observation of what is living and as a questioning of the nature of reality itself. For the Greeks, the soul is not merely the architrave of biological life but the fulcrum around which the quality and value of human existence revolve, a point of convergence between the sensible and intelligible worlds. Aristotle regarded life as an emanation of the final cause, an ordering principle, and wrote that *nature does nothing in vain*. An order is not rigidly mechanical but organic and dynamic. It evolves, grows, and adapts, even if we do not know where it takes us and why it develops. Thus, through philosophical wisdom and the call of myth, life manifests itself not as a mere accident but as a *founding* principle, an expression of the *logos* that continues to challenge, explore, and create in the vast theater of the universe.

In Hindu cosmology, the universe is permeated by the three Gunas – essential qualities that govern the nature of existence. *Tamas* (तमस्), darkness, is the force of inertia and ignorance, described in the *Bhagavad Gita* as follows: "Know that darkness (tamas) is born of ignorance and confounds all embodied beings" (Bhagavad Gita, 14.8). This Guna is analogous to the concept of entropy in physics; it represents the growing disorder, the inexorable flow toward chaos that threatens to engulf information and order. However, in Gunas' cosmic dance, there is more than just tamas. *Sattva* (सत्त्व), purity, is the quality of truth and enlightenment evoked in the *Mahabharata*: "When the light of wisdom spreads through all the gates of the body" (Mahabharata, 6.33.7). Such is the life-like force that ceaselessly reaches out toward complexity and order, opposing entropic dissolution. Sattva is the flame of awareness burning in the heart of Tama's darkness, a beacon of harmony in a sea of disorder. *Rajas* (रजस्), energy, is the quality of action and movement, which is described in the *Bhagavad Gita* as follows: "Know that rajas is passion, born of desire and attachment" (Bhagavad Gita, 14.7). Rajas is the intermediate drive that catalyzes and transforms, an *agent of change* that interacts with sattva and tamas, shaping the dynamic flow of existence. The three gunas are intertwined waves in the vast field of the universe, symbols of the fundamental forces in an eternal struggle with each other. In their interaction, the gunas embody the cosmic drama of existence, a symbolic representation of the tension between order and chaos, light and darkness, the battlefield on which life fights every day to assert its harmony against the background noise of entropy.

Thus, life becomes *the agent of coherence* that transforms energy into complex and meaningful structures, information, and consciousness. Life – understood as the movement of coherence in the universe – weaves energy plots and makes living and meaningful structures out of it in a growing architecture of information and consciousness. In his dialectical vision, Hegel saw human history as the stage on which the self-consciousness of the Spirit is unveiled. "Das Wahre ist das Ganze" ("The true is the whole"), he exclaims in his philosophical system, where every single event in history is an embodiment of the absolute Spirit proceeding toward self-realization through the tension and synthesis of opposites. In every act of learning and creation, the human witnesses and participates in his own evolutionary process. Life thus becomes an enterprise of deciphering and manifesting the inherent order. Humanity, in its striving for knowledge, is a reflection of that Hegelian idea that *becomes history*, that *becomes life*, evolving toward an ever more refined understanding of reality. Self-consciousness emerges as the result of a long journey that, according to Hegel, is prefigured in the very structure of Being. "Die Weltgeschichte ist der Fortschritt im Bewusstsein der Freiheit" ("The history of the world is the progress in the consciousness of freedom"), he stresses in his *Philosophy of History*, suggesting how each epoch, each society, each individual contributes to the revelation of an ever broader and deeper freedom. Thus, in the Hegelian interpretation, life is a dynamic unfolding of the absolute, a making of relations and meanings that *eschews* the irreversible arrow of *thermodynamic time* and stands as the protagonist of a *historical time* that becomes a web of freedom and awareness. In the dialogue between the finiteness of our condition and the infinitude of Spirit, human life is called to be not a mere witness but a true co-creator, in its modest but essential contribution, of the cosmic work of the evolution of Spirit itself.

In a universe destined for dissipation and thermal death, life emerges as a miraculous epiphenomenon, a rebellion against the current, a light in the darkness of ignorance and chaos. The consciousness that emerges is the spark of its transformation. But is a spark igniting in the universe enough to keep the darkness of the eternal at bay? Our existence is a thread strung between two extremes, a path that oscillates between the order created by our consciousness and the inevitable disorder that surrounds us. It leads us to ponder the theory of the universe as a kind of *cosmic mind*, an idea that some philosophers and scientists have explored, suggesting that the universe may have an *intrinsic* consciousness or *purpose* beyond known physical laws.

Life is thus not reduced to a mere biological phenomenon but becomes an act of (re)balance and challenge, *an attempt to imprint meaning and purpose into the matrix of time and space,* with its ability to create order from disorder, complexity from simplicity, questioning itself as a philosophical conundrum that invites more profound reflection on the meaning of existence and the laws that govern the cosmos. Such an enigma leads us to contemplate how life can exist in a universe governed by thermodynamics and its role in the larger order of existence, an order that might go beyond physical laws and touch the depths of the mystery of being.

From the simplest bacterium to the vast complexity of human society, life does not just resist entropy – it uses it, harnessing the constant flow of energy to create intricate structures, process information, grow, adapt, and survive. How can we learn and be inspired by this resilience? The universe is thus the grand setting for a *challenge to entropy* - a dialectic of successive creation and dissolution, a testament to the potential hidden in the folds of natural laws and a will that seeks to unfold it. The book sets out to explore this notion from different scientific and philosophical perspectives. The challenge is a search for the deep interaction between life and the universe that investigates our existence in all its bewildering complexity. It invites readers to look at the world around them and the universe within which it exists with a sense of wonder and a fundamental search for meaning.

From the Big Bang to Consciousness

From the Big Bang to Consciousness

From the Big Bang to Consciousness

PART ONE

THE FOUNDATIONS

From the Big Bang to Consciousness

From the Big Bang to Consciousness

> πάντα ῥεῖ
> Everything flows
>
> HERACLITUS (535 - 475 B.C.)

> In the midst of chaos,
> there is also opportunity.
>
> SUN TZU (544 - 496 B.C.)

From the Big Bang to Consciousness

From the Big Bang to Consciousness

Chapter 1

THE LAWS OF THE UNIVERSE

From the Big Bang to Consciousness

From the Big Bang to Consciousness

Two things fill the soul with admiration
And ever new and growing veneration,
the more often and the longer the reflection deals with them:
The starry sky above me, and the moral law in me.
These two things I don't need to look them up.
And simply assume them as if they were shrouded in darkness,
Or were they in the transcendent, outside my horizon;
I see them in front of me and connect them immediately
With the consciousness of my existence.

IMMANUEL KANT

Critique of Practical Reason
(Kritik der praktischen Vernunft), Conclusion, 1788

From the Big Bang to Consciousness

1.1 Laws of the Universe

For Kant, what is astonishing about the *physical* universe is that even in all its vastness and complexity, it is governed by fundamental laws intelligible to us. What is extraordinary about the *inner* universe is how a being as limited as the human aspires to an understanding of the Absolute. Have we ever stopped to ask ourselves whether we can *really* understand the whole, outside and inside us? The human, a *living* entity composed of two-thirds hydrogen atoms – the same elements formed just 3 minutes after the Big Bang singularity – finds himself questioning the laws of the same universe that gave birth to him more than 13.7 billion years ago. What, after all, is such a quest, if not a cry addressed to the Origin and the Creator himself? The universe, it is undeniable, through us becomes *self-aware*. But what does it really mean?

When we come across the laws that govern the Universe, it is fascinating to retrace the stories of those who discovered them and how a flash of genius moves our human race forward. Take, for example, Isaac Newton, the father of the law of universal gravitation. The famous apple falling from the tree is undoubtedly an anecdote. Still, it also represents a pivotal moment in the history of science, a moment when everyday observation stimulates human genius. An apple is falling – a law of the universe unraveling. The laws that seem to govern every corner of our vast universe, from giant stars to tiny particles, have been revealed through the curiosity and tenacity of scientists who have dedicated their lives to understanding the deepest mysteries. Reflecting on their journey and discoveries brings us closer to science and inspires us to pursue our own curiosity and search for the answers hidden in the apparent surface of reality.

In the myth of Prometheus, he who stole fire from the gods to give it to men echoes the innate curiosity and irrepressible desire to unravel the mysteries hidden in the universe's womb. The Promethean spark ignites the reaction in the face of the unknown, the primordial need to unveil what lies hidden in the abyss of reality. In the arc of his finite destiny, the human challenges the infinite, tending toward that fire that is a symbol of knowledge, light that makes its way, and clarity in the dark heart of the cosmos. Knowledge is light; the mystery is its shadow. In all philosophical and scientific exploration, humankind becomes the bearer of the fire that illuminates the darkest recesses of Being, seeking to transfigure mystery into knowledge and shadow into light. In its mythical unfolding, the universe is not a mere web but rather a living organism, a pulsating totality whose existence is saturated with hidden meanings and laws still hidden from our gaze. Science and philosophy, in their titanic impetus, rise to challenge the limits imposed by the human horizon in an effort to transcend phenomenal appearance and touch the ultimate essence of things, to answer the supreme question of man's location in the infinite arcane of the universe.

The laws of thermodynamics, the principles of quantum mechanics, are not just mathematical codes but rather the *very language of the universe,* an *original language* just waiting to be translated and, perhaps, one day, spoken fluently. At the heart of the myth lies the key to deciphering the enigma between life's essence and the universe's unfathomable laws. An enigma that winds between myth and reality, between the sacred and the profane, between the immanent and the transcendent, and which continues to trace the life of the cosmos, a design just waiting to be revealed. The living creature born from stardust now ventures into the mysteries of the cosmos, trying to decipher its arcane laws. The universe discovers itself through the laws of physics and chemistry, the fundamental principles that explain the interactions between matter and energy, from the smallest subatomic particles to the largest cosmic structures. Laws that encompass classical mechanics, thermodynamics, electromagnetism, relativity, and quantum mechanics and that guide the formation of stars, planets' orbits, and galaxies' intricate movement. But the laws are not just about macroscopic quantities. They govern even the smallest scales, from the chemical reactions that take place in every cell of our bodies to the interactions of subatomic particles, to the infinitely small, where space and time cease to exist and where perhaps the true face of the universe is hidden.

The introduction of *quantum mechanics* has highlighted the probabilistic and nondeterministic nature of the universe, giving rise to paradoxical aspects that cannot be reconciled with classical physics. What does this veil of uncertainty conceal? The probabilistic nature of quantum mechanics indicates that there are limits to our ability to predict future events accurately, that reality is *inherently indeterminate*, or even that there may be a *deeper order* underlying the classical laws of physics. Are we on the threshold of a new understanding or on the brink of an abyss of mysteries? There may be, as yet, unknown laws or principles that manifest themselves in particular circumstances or scales. The path of science is paved with questions.

Among the fundamental laws of classical mechanics are the *laws of thermodynamics*. The term usually refers to the four laws of thermodynamics that describe the behavior and relationships of transfers and transformations of energy, heat, work, and the property known as *entropy*, from the Greek word meaning *turn toward* or *transformation*. Change is the only constant. The laws concerning energy transfer apply as much to a star's core as to a living organism's metabolic processes, embracing both the microcosm and the macrocosm. At first glance, thermodynamics seems to dictate a gloomy fate for all things. It speaks of a universe that tends toward disorder, of energy that dissipates, of systems that break down. But is this really the case? And where would the inevitable tide of chaos take us?

Entropy, in a basic sense, can be considered a measure of disorder or randomness within a system. Could there instead be a hidden order behind the chaos? And why does the phenomenon of life apparently *seem to* defy such a law? Indeed, life *creates order from disorder* and complexity from simplicity. It harnesses energy to build intricate structures, grow, reproduce, and evolve. Is it chance or intentional design? Is it a rebellion or a form of submission to the universe's seemingly inevitable march toward entropy? How can life exist in a universe governed by thermodynamics? How can it defy the fate that the laws of the universe seem to dictate? One must examine the phenomenon in detail. Although life introduces some degree of order, it is not the only phenomenon that does so. Crystals, weather patterns, geological formations, and ordered movements of galaxies are all examples of how order can emerge in nature, even in the absence of biological processes.

The paradox of life in an entropic universe is often known as the *entropy paradox* or *Schrödinger's paradox*, named after physicist Erwin Schrödinger, who in his seminal 1944 book, *What is Life? (What is Life?)*, addresses the problem. He argues that life avoids decay toward equilibrium – the increase in entropy – by feeding on *negative entropy*, importing order from the environment through nutrients and sunlight(a concept he corrected in later stages of his speculation). Does life really defy the second law of thermodynamics? Or is it simply energy creating *temporary* local order at the expense of an overall increase in entropy in the universe? This is the great *mystery* we will explore in the book. We will delve into the laws of the universe, the principles of thermodynamics, and the nature of life itself. We will explore how life can exist in a universe governed by entropy and how it can even be a fundamental part of the universe's grand design.

The idea that life may be a fundamental part of a larger design of the universe is a topic often present in ongoing philosophical and scientific debate. It has led to the formulation of various hypotheses and theories, such as the *Anthropic Principle*, the *Gaia Hypothesis*, and the concept of the *Self-organizing Universe*. Although offering interesting *teleological* perspectives, such theories do not provide definitive conclusions about the nature of the universe or the role of life within it, continuing to be explored and refined by scientists and philosophers. Our understanding of life and the universe continues to evolve, and each discovery may rewrite or refine what we think we know.

1.2 Understanding thermodynamics

Thermodynamics is a fundamental branch of physics that deals with the relationship between heat and other forms of energy. It provides a quantitative description of how energy is transferred in natural processes and establishes the limits of what can and cannot occur in a closed system. But what does energy transfer or conservation mean in practical life? Let us pause for a moment to consider what this principle means in our daily lives.

With its intricate concepts, thermodynamics may seem like an abstract subject that is difficult to grasp, yet we can approach fundamental science through simple everyday analogies. Imagine, for example, that energy is like water in a river. Just as water flows from high points to low points, energy moves from higher concentration to lower concentration states, seeking balance. Just as a river can move a water wheel, energy, in its flow, can do work, powering the processes of life and man-made machines. Just think of when we heat water to make a cup of tea and then later wait for it to cool to drink. The analogies of energy in transformation help us visualize the concept of *energy flow*, a fundamental pillar of thermodynamics, and understand how, despite its invisible nature, energy is a powerful and omnipresent force that shapes our world.

One of the cardinal principles of thermodynamics is energy conservation, encapsulated in the *First Law of Thermodynamics*.

The First Law of Thermodynamics can be formulated as:

$$\Delta U = Q - W$$

Where:

U = Internal energy of a system
Q = Heat added to the system
W = Work done by the system

The simple and powerful first law equation states that the change in internal energy ΔU of a system equals the amount of heat Q added to the system minus the work W done by the system on its surroundings. The law encapsulates the eternal cycle or the *principle of conservation* of energy. It is a principle of balance and harmony that governs every aspect of the universe.

And then there is entropy.

A concept that often eludes and defies intuition. It is chaos set against order, a surprise in a seemingly orderly universe. An essential pillar of thermodynamics is the *Second Law*, often described as the measure of disorder in a system.

The Second Law of Thermodynamics can be formulated as:

$$\Delta S \geq Q/T$$

Where:

S = Entropy
Q = Heat transferred
T = Absolute temperature (measured in Kelvin)

The law suggests that the entropy of an *isolated system* will always increase over time and that heat cannot spontaneously flow from a colder to a warmer place. Perfectly isolated systems do not exist, but the concept is helpful for theoretical discussions. However, life apparently contradicts the second law of thermodynamics by organizing itself into highly ordered structures. The book explores precisely the idea that life, in its complexity, might be the universe's way of coping with increasing entropy. On the other hand, let's try to imagine a universe where entropy does not increase. What would it be like? Static, unchanging, dead? Or perhaps a universe where time ceases to exist and crystallizes into an eternal present?

The *Third Law* of Thermodynamics concerns absolute zero, the lowest possible temperature at which all molecular motion ceases. It states that when a system's temperature approaches absolute zero, its entropy approaches a minimum or constant value.

$$\text{When } T \to 0, S \to \text{a constant.}$$

It represents the only case where entropy is zero, assuming a perfect crystal structure. In absolute zero, entropy is revealed in its most enigmatic guise, perhaps the most profound key to understanding the relationship between energy and matter.

Thermodynamics also introduces us to the concept of *free energy*, *Gibbs free energy* (G), a measure of the maximum reversible work that a thermodynamic system can perform at constant temperature and pressure, and *Helmholtz free energy* (A), which are useful for systems at constant temperature and pressure and for systems at constant temperature, respectively.

Such fundamental laws and concepts serve as the basis for exploring the flow of energy in the universe, the origins and structure of life, the workings of the human mind, and much more. They lay the foundation for understanding the universe's behavior at the macroscopic level, establishing how energy moves and changes, how heat affects matter, and how work takes place. They are fundamental to our understanding of the universe and everything around it. However, when we observe life, we see a system seemingly defying this law. Life organizes itself into highly ordered structures, maintains its complexity, and even increases it over time through evolution. The book's central theme is the apparent contradiction between the second law of thermodynamics and the complex order of life.

In the following sections, we will delve into the second law and the concept of entropy, explore the apparent anomaly of life, and introduce the concept of *dissipative structures*. Ilya Prigogine, who won the Nobel Prize in Chemistry in 1977 for his work on *non-equilibrium thermodynamics*, particularly the theory of dissipative structures, popularized the term. Examples include cells, which use energy to maintain structure and function, and convective cells in the atmosphere, which form when the sun heats the Earth.

1.3 The Second Law and the mystery of the origin of the universe

The Second Law of Thermodynamics is considered one of the inescapable laws of the universe and rests on a foundation that is not yet fully understood. What does this law imply for scientific observation and our everyday life? It introduces the concept of entropy, a measure of disorder or randomness in a system. The law states that the entropy of an isolated system will always increase over time or remain unchanged in the ideal cases of reversible processes. And is not life itself a reversible process, at least in part? It is often interpreted as the natural tendency of the universe to disorder. The concept, while abstract, is tangibly manifested in our daily lives. Let us pause for a moment. Let us think, for example, of how a tidy room becomes messy over time (especially our children's!) without outside intervention (the work of tidying up). Inevitable, isn't it? Or how the heat from a cup of coffee dissipates into the environment until thermal equilibrium is achieved. Can the simplicity of everyday observations teach us something more profound about the universe?

Following the most scientifically plausible hypothesis, the Second Law of Thermodynamics inevitably takes us back to a singularity, a starting state *T₀ where the universe was in a state of infinite order*. This is one of the most challenging puzzles for modern physics to grasp. How could a universe, which now manifests an irrevocable march toward greater disorder, have begun from a state of perfect order? Could perfect order be just an abstract concept, a cosmic mirage?

The enigma takes us on a journey through cosmic time, from the point of singularity through the Big Bang to the evolution of the cosmos as we know it. It is a journey that begins with a question: where did the initial order come from? The origin from a state of infinite order implies a condition of little or no entropy, where energy is concentrated, not dispersed. Space-time begins with precise, measured motion. As the universe expanded, energy then spread out, entropy increased, and complexity emerged in the most surprising ways, not least in the form of life. Complexity from simplicity: isn't that the real magic? The notion that entropy was zero at the time raises profound questions about the nature of the universe in terms of the *infinitely small probability of initial conditions*. How can chance play a crucial role in a finely synchronized universe? A null entropy condition suggests extreme order and specificity at the initial instant of the universe, which challenges our current understanding of how time, space, and matter developed. What if the initial order is the key to unlocking the following mysteries? From a probabilistic perspective, it implies a singular state of high specificity, contrasting with the concept of entropy as a measure of disorder. Extreme disorder, then, is not only the end but also the beginning.

The resolution of this paradox remains one of the great mysteries of contemporary physics. It underscores the need for an interdisciplinary approach that unites physics, mathematics, and philosophy to decipher the fundamental puzzles of the universe. The lens of interdisciplinarity is imperative if clarity is to be achieved. One theoretical explanation might be related to the extreme conditions and physical laws operating in the primordial universe. But how do the physical laws behave when the universe itself is incubating? Some cosmological theories, such as the *cosmic inflation* theory, suggest that the early universe was infinitely hot, dense, and uniform at the instant of the Big Bang. Uniformity could be interpreted as the highest state of physical order. But how does uniformity preserve the secret of the whole? Our reasoning mind leads us to question the conditions before the initial instant, perhaps speculating on the end of a previous universe state.

Moreover, gravity plays a crucial role in such a scenario. Under conditions of high density, such as those in the primordial universe, gravity might have acted to maintain an infinite degree of order, counteracting the increase in entropy observed under less extreme conditions. Gravity could be the force that counterbalances entropy by substituting order for disorder.

Quantum field theory and particle physics offer further explanations. Under such extreme conditions, quantum fluctuations and interactions of elementary particles may have generated an initial order that then gave rise to the observed complexity in the present universe. We could be, in a sense, children of the quantum vacuum fluctuations. A fundamental point is the very notion of *time* in relation to entropy. What if time is just a driving force rather than a mere measurement? If time itself emerges with the Big Bang, our understanding of entropy as something that increases *over time* may not be applicable in the context of a pre-Big Bang universe. Explaining an initial moment of maximum order and low entropy in the primordial universe remains one of modern physics's most fascinating open questions. The answers may require new physics or a revision of our current cosmological and thermodynamic theories. They could lead to a deeper understanding of the universe's nature and its fundamental laws.

Philosophically, the idea of a universe emerging from an infinitely small or even a-dimensional point – a *cosmic egg* of perfect order moving toward a condition of increasing disorder – embodies a narrative of loss and rebirth, destruction and creation. Does it reflect the principle of the eternal cycle of end and new beginning? Or does it reflect the concept of *kenosis* – the sacrificial withdrawal of God as conceived by the Church Fathers of the Eastern Christian tradition – or *tsim-tsum* – the withdrawal of God according to Jewish mysticism to make room for his creation? Staying in a more immanent dimension, life could manifest the universe's intrinsic desire to explore all possible configurations to express itself in myriad complex ways while still within the boundaries imposed by spacetime.

The mystery of the universe's ordered origin and subsequent march toward disorder, with life dancing in the midst of it, invites us to reflect not only on the physical laws that govern the cosmos but also on the deeper meaning of existence, why order and disorder exist, and the unique place that life occupies in this grand development. The dialogue between the universe's ordered past and its disordered present and future, with life bridging the two extremes, remains one of the deepest mysteries of physics and philosophy. This mystery continues to challenge our understanding and inspire our wonder.

Entropy is a difficult concept to grasp because it is counterintuitive to our everyday experiences. But is it not in paradoxes that we find the greatest enlightenment? We see seeds growing into complex and ordered structures like trees. We witness the development of a human being from a single cell to a highly organized being. We observe cities and societies, with their intricate systems and structures, emerge from seemingly chaotic conditions. Phenomena that seem to contradict the Second Law since they involve an increase in order and complexity, not disorder. Is this perhaps an illusion, or are they two sides of the same coin?

It is crucial to understand that the Second Law applies to isolated systems that do not exchange matter or energy with their surroundings. Earth, for example, is not an isolated system. It constantly receives energy from the Sun, which powers many processes on our planet, including life itself. The Second Law, in fact, does not prohibit local decreases in entropy. While the overall entropy of an isolated system cannot decrease, some parts of the system can become more orderly at the expense of increased disorder in other parts of the system. This is how life can exist and create order in a universe that, as a whole, tends toward disorder. Perhaps it is in the balancing of the cosmic double-entry that life finds its stage. Entropy, however, is not only about disorder. It is also about energy dispersion and the natural number of ways a system can be organized. In this sense, disorder probabilistically reflects the universe's maximum freedom of matter development. A system naturally evolves toward the state with *the greatest number of possible arrangements*, the state with the highest entropy. For there are simply more ways of being disordered than ordered. Doesn't natural evolution prompt us to ask what *natural* really means? In the context of life, the Second Law and entropy play a crucial role. As we will see in the following sections, life is a process that harnesses energy to create order and complexity, seemingly defying the march toward disorder. It is fascinating to think that every life form is a rebel born within the empire of entropy.

The Second Law introduces the concept of *irreversibility* or the *arrow of time*. It implies that some physical (and biological) processes can only occur in one temporal direction, which is why we perceive that time moves only forward. At the macroscopic level, time and entropy could be one epiphenomenon of the other. For example, an ice cube melts in water if left in a warm room, but a tub of water does not spontaneously form ice cubes at room temperature. What we observe as simple results from complex, invisible rules. What does this teach us about the flow of our own existence? The Law establishes a fundamental limit on efficiency or limit on *energy conversion*. Only some of the energy a heat engine supplies can be converted into work; some of it will always be *wasted* in the form of heat. This principle underlies the operation of all engines and power plants. Could it contain a more profound message, a lesson about the inherent nature of loss and gain? The Second Law also helps us determine whether or not a process can occur spontaneously (*spontaneity of processes*). A process will be definable as spontaneous if it leads to an increase in the total entropy of the universe and reveals to us the direction in which the universe prefers to move.

The principles of the Second Law are fundamental to understanding how structures are formed, including life itself (*formation of structures*). Despite the law's assertion of increased disorder, it also allows for localized order, such as the complex structures of life, as long as it is accompanied by increased disorder elsewhere in the universe. Whenever we admire the beauty of a flower, are we observing the balance between order and chaos, the universal account balancing before our eyes?

1.4 Life: an anomaly?

Life seems to defy the natural tendency toward disorder, creating intricate structures and systems from simple building blocks. An enigma that invites more profound reflections. From the single-celled organisms that first populated our planet to the vast species that inhabit it today, life displays an extraordinary capacity for order and complexity. How does life emerge from the apparent silence of cosmic matter? On the one hand, life seems to defy the relentless increase in entropy, creating order from disorder, as we see in the complexity of living organisms. Might not what we see around us be evidence of a hidden force driving the universe? Scientists mostly agree that, in fact, life does not contradict the Second Law of Thermodynamics but rather is a manifestation of it. On the contrary, they argue that life, through its metabolic processes and from a materialistic point of view, actually contributes to the overall increase in the universe's entropy. The debate opens a fascinating window into the deep questions concerning the nature of life and its place in the cosmic order.

We regard the human body as a marvel of biological engineering, (perhaps) the pinnacle of the current natural order. It is composed of trillions of cells, each a microcosm of complexity, performing a myriad of functions that sustain us. Our bodies can grow, repair, and reproduce while maintaining a high degree of order. This level of organization, of complexity, seems to counter the universe's apparent march toward disorder. Or consider an ecosystem, a complex network of interactions between various organisms and their environment. Is this not an example of perfect, almost musical harmony? Systems exhibit a high degree of order and balance, with energy flowing from the Sun through plants and the food chain, driving a cycle of life, death, and renewal. Again, this seems to contradict the dictate of the Second Law on the increase of entropy.

But is life just an anomaly, a lucky accident in a universe that tends toward disorder? Or could it be something more? Could life be the means by which the universe explores itself, experimenting with form and function? Could life be a fundamental part of the universe's strategy to increase entropy? The question of life as a possible anomaly or as an integral part of the cosmic order has intrigued minds through the ages. Many traditions see life as a manifestation of *divine will* or a higher order. In the Christian tradition, for example, life is seen as a divine creation, an emanation of God's love and wisdom. Offering a perspective in contrast to the view of a universe marching inexorably toward disorder suggests that perhaps every atom, every cell, holds a *breath* of divine intent.

Mythologically, life is often seen as a force emerging from primordial chaos. In Greek mythology, for example, life emerges from Chaos, a primordial condition of disorder, through a series of divine generations leading to the known world's order and structure. It is a myth that reflects a deep understanding of the tension between order and disorder, a tension that life seems to navigate and transform, a tension that we experience every day. Is it not precisely the eternal struggle between chaos and cosmos that not only every ancient mythology but also Greek tragedy up to modern psychoanalysis tells us about?

Life can be seen as a manifestation of the impulse of being toward expression and complexity. Alfred North Whitehead suggested that life reflects a tendency of the universe toward innovation and creativity, which is expressed through the evolution of biological complexity. In this sense, life is not an anomaly but the manifestation of a *deep cosmic tendency toward self-organization* and emergent complexity. The question of whether life is a mere accident or a fundamental aspect of the cosmic order touches the heart of our understanding of the universe and our place in it. We, therefore, face an epistemological crossroads: which path to choose to explain the existence of such a powerful and pervasive force? The fundamental question invites us to explore not only the laws of physics but also the profound philosophical, religious, and mythological implications of life in an evolving universe. Through the lenses of religion, myth, and philosophy, life is revealed not as an anomaly but as a profound and intrinsic expression of the order and creativity of the universe.

1.5 The concept of dissipative structures

To reconcile life's existence with the Second Law of Thermodynamics, we must introduce the concept of dissipative structures. Belgian physical chemist Ilya Prigogine coined the term, which refers to systems that maintain or increase their structure and complexity by exchanging matter and energy with their environment. This raises an essential question: Is order really a gift or a temporary borrowing from the universe? The concept is a key to understanding how seemingly ordered systems can emerge in a universe that tends toward disorder.

As Prigogine explains, dissipative structures are *open systems* that maintain their state of complexity and order through the continuous exchange of energy and matter with the environment. A prime example is living cells, which maintain their complex structure through metabolic processes that exchange energy with the outside world. Yet another example can be observed in the eddies that form in a moving liquid, where the energy embedded in the system maintains a temporary order amidst the chaos.

Let's pause for a moment to contemplate the beauty of temporary perfection. A vortex is formed in a body of water when energy (in the form of water flow) is introduced into the system. The vortex maintains its structure and complexity – its vortex pattern – by continuously dissipating energy in its environment. If the energy flow is interrupted, the vortex dissipates, and the system returns to equilibrium. How often have we observed nature's return to equilibrium, a reminder of how nothing is permanent? Such examples help us visualize how order can arise from disorder and how dissipative structures are a fascinating and vital manifestation of the thermodynamic laws at work. Dissipative structures are systems far from equilibrium; that is, they exist in a state of constant flow maintained by the continuous flow of energy through the system. The flow of energy allows the system to maintain its structure and increase its complexity over time despite the general tendency of the universe toward disorder.

Life, in many ways, can be considered a dissipative structure. Or perhaps. Life may be the most majestic of dissipative structures. Organisms maintain their complex structures by continuously consuming energy (in the form of food) and dissipating it into the environment (in the form of heat and waste). The continuous flow of energy allows life to maintain and even increase its complexity over time, despite the dictate of the Second Law about increasing entropy. In this context, life does not defy the Second Law but embraces it. By creating local order and complexity, life increases the overall entropy of the universe. It is a process that harnesses energy to create structure, grow, reproduce, and evolve while driving the universe's march toward disorder.

The concept of dissipative structures takes us into territory where physics meets metaphors that resonate through deeper dimensions of understanding. It is no accident that the histories of humanity have always played with the themes of ebb and flow, of construction and dissolution.? Even the ancient interplay between *Yin* and *Yang*, between being and non-being, as a dissipative structure, seems to reflect life. These structures can be seen as a tangible manifestation of the dynamic, the heart of many spiritual traditions, between order and chaos. For example, in Daoism, the concept of *Dao*, ultimate harmony and order, is in a continuous balancing act with the dynamic chaos of the manifested world. As a dissipative structure, life reflects the cosmic duality, maintaining an ordered form through the constant flow of energy.

The same concept also invites reflection on the nature of the divine. Could divinity itself be understood as the architect of an eternal change process, the driving force behind all exchange and transformation? Could divinity be seen as the flow of vital energy that enables the manifestation of dissipative structures? This invisible breath expresses itself in the complexity of the world around us, from the swirl of water to the texture of a leaf. Could the divine energy, flowing through and animating creation and manifesting itself in every dissipative structure, reflect the *divine desire* for exploration, expression, and experience? Are we not, all of us, the explorers of the universe seeking to understand the flow of energy that unites us?

1.6 Life as a dissipative structure

In all its myriad forms, life can be seen as a complex network of dissipative structures. But what exactly does a dissipative structure mean? From the simplest single-celled organisms to the most complex ecosystems, life is able to maintain and significantly increase its structure and complexity in the face of the universe's general trend toward disorder. Consider a single cell, the basic unit of life. A cell is a tiny factory of complexity, a microcosm of ingenuity, continuously taking in nutrients from the environment, converting them into energy through metabolic processes, and dispersing waste products into the environment. A balancing act, a continuous flow of matter and energy, allows the cell to maintain its complex structure and perform the functions necessary for life in miniature. How can something so tiny be extraordinarily complex at the same time? To better visualize the concept, we can imagine a cell as a dynamic factory in which raw materials (nutrients) enter and wastes (products of metabolism) leave, thus keeping the cell in a state of order despite the surrounding entropy. It is no accident that the cell, in its complexity and efficiency, represents the very basis of life.

On a larger scale, consider a plant. A plant absorbs sunlight, carbon dioxide, and water from its environment. It converts them into glucose and oxygen through *photosynthesis*, the process by which green plants and other organisms use sunlight to synthesize food with the help of chlorophyll pigments. Oxygen – the miracle that sustains us with every breath – is a byproduct of that process. Energy-rich glucose fuels plant growth and reproduction, while oxygen is released into the environment. Again, we see a continuous flow of matter and energy that enables the plant to maintain and increase its complexity for life to sustain other life.

On an even larger scale, consider an ecosystem, a complex network of interactions between various organisms and their environment. Energy flows through the ecosystem from the Sun through plants and the food chain, resulting in a cycle of life, death, and renewal. Despite nature's apparent disorder, ecosystems exhibit a high degree of order and balance, maintained by the continuous flow of energy.

In such examples, we see life acting as a dissipative structure, which maintains and increases complexity by continuously exchanging matter and energy with its environment in an endless cycle. Viewing life as a dissipative structure offers a surprising perspective on our existence in the universe. Such a viewpoint sees living organisms, including humans, as complex systems that maintain and develop their structure through a constant flow of energy and matter.

But what lies behind pure science? The idea of life as a dissipative structure can be seen as a manifestation of the divine will or the immanent creative principle that permeates the universe. It perhaps invites us to look further and seek more profound meaning, for example, in the Egyptian creation myth, where the god *Atum* emerges from the primordial watery chaos (*Nun*) and creates order through his thoughts and words. The mythological image reflects how life, as a dissipative structure, emerges, maintains, and evolves through the continuous flow of energy and matter, creating order from disorder.

What lessons can we learn from life's constant struggle? Life as a dissipative structure can be interpreted as an expression of the inherent potential of Being to manifest itself in increasingly complex and *teleologically* ordered (goal-directed) forms. Philosophers such as Teilhard de Chardin explored the idea that life and evolution are directed toward greater complexity and awareness, a process he called the *Omega Point*.

Through its dissipative nature, life can be seen as a vehicle through which the universe explores, expresses, and realizes its potentialities in myriad unique and complex forms, perhaps oriented toward a point of ultimate fulfillment. Isn't it amazing how, despite everything, life always finds a way? The concept of dissipative structures invites philosophical reflection on the idea of harmony and balance. Despite the general tendency of the universe toward disorder, life displays an extraordinary ability to create order, beauty, and harmony, defying the forces of disorder and dissipation. Therefore, the view of life as a dissipative structure enriches our scientific understanding without satiating our thirst for knowledge but opening up new, more profound existential, religious, and philosophical questions.

1.7 The role of energy flow

Energy flow is a fundamental aspect of life and its evolution. Have we ever stopped to think how deeply it touches every aspect of our existence? The role of energy flow in the universe, particularly in life on Earth, is a topic that profoundly touches our existence and future. It is the invisible golden thread that connects everything alive. Every organism, from the simplest algae to complex human systems, depends on a constant energy supply to survive and thrive. Energy flow is not just an abstract concept in physics; it is the heartbeat of nature and has direct and significant implications for our everyday actions. How often do we remember that every morsel of food we ingest is nothing more than transformed sunshine? In the ecological context, we can see how solar energy captured by plants through photosynthesis is the basis of the food chain, thus sustaining all life on Earth. Isn't it amazing that a ray of sunshine becomes our sustenance? At the same time, the growing awareness of the importance of renewable and sustainable energy sources highlights the need for careful management and use of energy flow. Such understanding prompts us to reflect on our environmental impact and the choices we make every day, emphasizing how respecting the natural laws of energy is crucial to the well-being of our planet and future generations.

The Sun is the primary source of energy for life on Earth. Through the silent transformation of photosynthesis, plants capture sunlight and convert it into chemical energy in the form of glucose. Its energy is then used to fuel plant growth and reproduction and is also passed down the food chain when other organisms consume plants. The continuous flow of energy from the Sun through plants and the food chain allows life to maintain and increase its complexity. It will enable life to exist as a dissipative structure, creating local pockets of order and complexity that increase the overall entropy of the universe. Organisms more capable of capturing and using energy from their environment are more likely to survive and reproduce. Over time, this leads to the evolution of increasingly complex and efficient organisms, and the flow of energy allows life to exist and drives its evolution.

In many religious traditions, *divine* energy or *light* is seen as the life force that pervades and sustains all of existence. Solar energy, which powers photosynthesis and maintains the food chain, could be seen as a physical manifestation of divine energy, an ongoing blessing through which the divine nourishes and sustains life on Earth. For example, in Hindu mythology, the sun god *Surya* is revered as the source of light and life, and his role in nurturing life on Earth mirrors the dynamics of energy flow in photosynthesis and the food chain.

1.8 Evolution: a response to entropy

Shouldn't we name the miracle of order emerging from chaos *evolution*? Evolution is the driving force of complexity and order observed in biological systems. It allows life to exist as a dissipative structure in a universe governed by the Second Law of Thermodynamics. Evolution can be seen as an expression of the *cosmic tendency toward the creation of new forms* and the *realization of latent potentialities*.

Evolution, the process by which life changes and diversifies over time, can be seen as a direct response to the Second Law of Thermodynamics. It is a testament to life's ability to harness energy flow and create complex and ordered systems in a universe that generally tends toward disorder in a revolt against the inevitable fate of decline. The process, which has shaped the diversity and complexity of the living world, seems almost a counterpoint to the universal tendency toward disorder. Such a view of evolution also gives us a fascinating perspective on the intrinsic relationship between biological processes and the fundamental physical principles of the universe.

Evolution is driven primarily by natural selection, a process in which organisms best adapted to their environment tend to survive and produce more offspring. At the same time, however, making more offspring seems to respond to the need to counteract the high mortality of offspring. A strategy, a numbers game against adverse odds, a race against time, a desperate attempt to leave a mark before the curtain falls? Over many generations, the process leads to changes in species characteristics and the emergence of new species. An endless metamorphosis. But how does it relate to thermodynamics and entropy? To understand this, we need to consider the role of energy in natural selection. Organisms that can more efficiently capture and use energy from their environment have a survival advantage: a greater chance of surviving, reproducing, and passing on their traits to the next generation.

The drive for energy efficiency leads to increasingly complex structures and systems. For example, the evolution of photosynthesis allowed plants to capture energy directly from the Sun, leading to a massive increase in the complexity and diversity of life on Earth. Similarly, the evolution of complex brains allowed some animals to navigate the environment better, find food, and avoid predators, leading to an increase in complexity. In this way, evolution can be seen as a process that increases the complexity and order of life, seemingly in contrast to the general tendency of the universe toward disorder. As we have already seen, the increase in complexity is driven by the flow of energy through a system and leads to a rise in the overall entropy of the universe.

Mythological stories of beings changing shape or acquiring new interpretive keys in response to the challenges they face in overcoming the adversities of the present are echoes of a past that *reflect on a symbolic level the evolutionary process of adapting to and overcoming environmental challenges.* The vision of an evolving universe could be seen as an insult to the concept of an unchanging natural order or, conversely, as a confirmation of the unceasing dynamic of becoming that many philosophers have identified as the heart of reality. As a response to entropy, evolution could be interpreted as a manifestation of matter and life's inherent potential for self-organization and self-transcendence. And in its extraordinary capacity for self-transcendence, might we not glimpse the image of a universe constantly renewing itself? The complexity and ingenuity of life could be interpreted as expressions of divine wisdom and intelligence, offering a vision of life as an integral part of a larger cosmic design. The idea that life could be the universe's way of creatively challenging entropy could lead to new reflections on the relationships between being, becoming, and the cosmic order.

Notes and insights to Chapter 1

1. Immanuel Kant (1724-1804), among the most important modern German philosophers, investigated the mechanisms of how the laws of the universe can be accessible to human understanding despite our finite and limited nature. For further study, see his (1781) *Critique of Pure Reason* (original title *Kritik der reinen Vernunft*) and (1788) *Critique of Practical Reason* (*Kritik der praktischen Vernunft*).

2. The Big Bang is the scientific theory that describes the universe's origin about 13.7 billion years ago from a singularity, a point of infinite density and extremely high temperature. See: Weinberg, S. (1977). *The First Three Minutes*.

3. The Prometheus myth, originating in Greek mythology, tells of Prometheus stealing fire from the gods and giving it to humanity, symbolizing the gift of knowledge. For more details, see Hesiod's works or Aeschylus's tragedies.

4. Quantum mechanics is a physical theory that describes the behavior of particles on a microscopic scale. It introduces concepts such as superposition and indeterminacy.

5. Laws of Thermodynamics: The four laws of thermodynamics govern the transfer and transformation of energy, heat, and work and introduce the concept of entropy.

6. Schrödinger's Paradox: Erwin Schrödinger, in his book (1944) *What is life? The Physical Aspect of the Living Cell*, explored the paradox of how life seems to defy increasing entropy by maintaining local order at the expense of an overall increase in entropy in the universe.

7. The Anthropic Principle suggests that the universe seems finely tuned to support life, often divided into weak and strong versions.

Reference: Barrow, J.D., Tipler, F.J. (1986). *The Anthropic Cosmological Principle*.

8. The Gaia Hypothesis proposes that the Earth functions as a self-regulating organism, with life playing a crucial role in maintaining stable environmental conditions. Reference: Lovelock, J. (1979). *Gaia: A new look at life on Earth*.

9. Various models and theories explore the idea that the universe may have self-organizing properties, suggesting that complexity may emerge from simple rules and local interactions. For more information, see Kauffman, S. (1996). *At Home in the Universe: The Search for the Laws of Self-Organization and Complexity*.

10. Dissipative Structures: the term, introduced by Ilya Prigogine, refers to systems far from equilibrium that maintain an ordered structure by dissipating energy into the environment. Reference: Prigogine, I., and Stengers, I. (1984) *Order Out of Chaos*.

11. Singularity, Big Bang, and the Origin of Time: The singularity, a starting point where the universe was in a state of infinite order, is a concept that ties in with the Big Bang theory. For more: S. Hawking (1988). *A Brief History of Time*. For discussion of the irreversible march toward greater disorder than the initial state of perfect order, one could cite the contributions of Roger Penrose, especially his concept of *asymmetric time* and the hypothesis of low initial entropy (Penrose, R. *The Road to Reality*, Vintage Books, 2004). The theory of cosmic inflation, which explains the uniformity and high initial temperature of the universe, has been extensively developed by Alan Guth and Andrei Linde (Guth, A. *The Inflationary Universe*, Addison-Wesley, 1997; Linde, A. *Particle Physics and Inflationary Cosmology*, Harwood, 1990). For the notion of time in relation to entropy, the works of Julian Barbour offer a perspective on the nature of time and its emergence with the Big Bang (Barbour, J. *The End of Time: The Next Revolution in Physics*, Oxford University Press, 1999). Still reflecting on time and its origin, I recommend the work of Carlo Rovelli, particularly *The Order of Time*, Adelphi, 2017.

12. Irreversibility and the Arrow of Time: The Second Law also introduces the concept of irreversibility, explaining why some processes can only occur in one direction in time. For more information, see: R. Penrose (2011). *Cycles of Time: An Extraordinary New View of the Universe*.

13. Energy Efficiency: The Second Law places a limit on energy conversion, a principle underlying the operation of motors and power plants. For further discussion, see: Cengel, Y.A., and Boles, M.A. (2006). *Thermodynamics: An Engineering Approach.*

14. Spontaneity of Processes: The Second Law helps determine whether a process can occur spontaneously, providing a framework for analyzing spontaneity in terms of increasing total entropy. For more information, see: D. Kondepudi, I. Prigogine (1998). *Modern Thermodynamics: From Heat Engines to Dissipative Structures.*

15. The concept of structure formation, including life, in relation to entropy and the Second Law, is a crucial area of investigation. To explore further, see: I. Prigogine (1980). *From Being to Becoming: Time and Complexity in the Physical Sciences.*

16. Ecosystems are complex networks of interactions between organisms and the environment, exhibiting a high degree of order and balance. For more details, see: E.P. Odum, G.W. Barrett (2004). *Fundamentals of Ecology.*

17. The relationship between life and increasing entropy is a fascinating area of investigation. Life could be a means by which the universe increases entropy. For more details, see: I. Prigogine (1997). *The End of Certainty.*

18. Different traditions offer unique interpretations of the role of life in the cosmic order. For further study, see: K. Armstrong (2005). *A Short History of Myth.*

19. Alfred North Whitehead explored life as a manifestation of a cosmic tendency toward self-organization and emergent complexity. For more details, see: A.N. Whitehead (1978). *Process and Reality.*

20. Life: anomaly or fundamental aspect of the cosmic order? The question touches on profound philosophical, religious, and scientific implications. Life may represent a manifestation of the order and creativity of the universe. For further details, see: P. Davies (1995). *Are We Alone?* Exploring the nature of life can enrich our understanding of the universe and our place in it. For further details, see: C. Sagan (1995). *Pale Blue Dot: A Vision of the Human Future in Space.*

21. Life can be seen as a complex manifestation of dissipative structures, maintaining and increasing its complexity through the consumption and dissipation of energy. For further details, see: E. Morowitz (1979). *Energy Flow in Biology*.

22. Dissipative structures do not defy the Second Law of Thermodynamics but are a manifestation of it, contributing to the overall increase in the universe's entropy. For more details, see: R. Swenson (1989). *Emergent Attractors and the Law of Maximum Entropy Production*.

23. The concept of dissipative structures resonates with ideas of order and chaos in spiritual traditions such as Daoism, reflecting the cosmic dynamic of harmony and energy flow. For more details, see: Laozi (1997). *Tao Te Ching*.

24. Dissipative structures also invite deeper reflections on the divine and the flow of vital energy that animates creation, offering a metaphor for exploring the interplay between order, chaos, and divinity. For further details, see: P. Davies (2000). *The Fifth Miracle*. The concept expands the understanding of order, disorder, and creativity in the context of physics, biology, and spirituality, providing a framework for exploring the emerging complexity and nature of the divine. For further details, see: F. Capra (1996). *The Web of Life*.

25. Life, manifesting itself in multiple forms, represents a complex network of dissipative structures. For more details, see: J. Schneider, D. Sagan (2005). *Into the Cool: Energy Flow, Thermodynamics, and Life*.

26. The analysis of the cell as the basic unit of life that maintains its structure through the continuous exchange of matter and energy with the environment. For further details, see: H. Morowitz (1979). *Energy Flow in Biology*.

27. The example of the plant that, through photosynthesis, maintains and increases its complexity by absorbing energy and matter from the environment. For more details, see: N. Lane (2015). *The Vital Question: Energy, Evolution, and the Origins of Complex Life*.

28. Ecosystems, with the flow of energy maintaining a high degree of order and balance, illustrate life as a dissipative structure on a larger scale. For more details, see: S. Levin (1998). *Ecosystems and the Biosphere as Complex Adaptive Systems*.

29. Mythological and religious interpretations of life as a dissipative structure in relation to the divine will or immanent creative principle. For further details, see: J. Haught (2000). *God After Darwin: A Theology of Evolution.*

30. Teilhard de Chardin's idea of life's evolution toward greater complexity and awareness explores the concept of the Omega Point. For further details, see P. Teilhard de Chardin (1955), *The Phenomenon of Man.*

31. The view of life as a dissipative structure opens a philosophical and existential reflection field, exploring themes of order, beauty, harmony, and the inherent potential of Being. For more details, see: T. Deacon (2012). *Incomplete Nature: How Mind Emerged from Matter.*

32. Energy flow is crucial to life and its evolution, driving the complexity and order observed in biological systems. For more details, see: Eric D. Schneider, Dorion Sagan (2005). *Into the Cool: Energy Flow, Thermodynamics, and Life.* Evolution, stimulated by energy flow, reflects the cosmic tendency toward the creation of new forms and the expression of latent potentialities. The importance of the Sun as the primary source of energy for life on Earth, fueling photosynthesis and the food chain. For more details, see: Govindjee, Beatty, Gest, Allen (2005). *Discoveries in Photosynthesis*; Raven, Peter H., Evert, Ray F., Eichhorn, Susan E. (2005). *Biology of Plants.* Evolution toward organisms increasingly able to capture and use energy from their environment. For further details, see: Dawkins, Richard (1996). *Climbing Mount Improbable.*

33. Religious interpretations of divine energy or light as a life force that permeates existence. For more details, see: Eliade, Mircea (1959). *The Sacred and the Profane: The Nature of Religion.* The mythological interpretation of energy flows through the adoration of the sun god Surya in Hinduism. For further details, see: Klostermaier, Klaus K. (1984). *Mythologies and Philosophies of Salvation in the Theistic Traditions of India.*

34. The role of energy efficiency in natural selection and evolution. For further details, see: Dawkins, R. (1976). *The Selfish Gene.*

35. Philosophical interpretations of evolution and the dynamics of becoming. For further details, see: Bergson, H. (1911). *Creative Evolution.* Evolution as a manifestation of self-organization and self-transcendence. For further information, see: Kauffman, S. (1993).

The Origins of Order. The increase in complexity and order through evolution as a response to the universe's tendency toward disorder. For more details, see: Carroll, S.B. (2013). *Brave Genius.*

From the Big Bang to Consciousness

From the Big Bang to Consciousness

Chapter 2

THE ORIGIN OF LIFE

From the Big Bang to Consciousness

From the Big Bang to Consciousness

> An honest man,
> armed with all the knowledge available to us now,
> could only state that in some sense,
> the origin of life appears at the moment to be almost a miracle,
> so many are the conditions which would have had
> to have been satisfied to get it going.
>
> Francis Crick
>
> *The Origin of Life*, 1981

From the Big Bang to Consciousness

2.1 The primordial soup

In the vast expanse of the cosmos, in an outlying galaxy, on a small blue planet orbiting an average star, life emerged – a spark in a cosmic abyss of infinity. The birth of life, a miraculous event, is one of the deepest mysteries that science – and not only science – has always tried to unravel. As the Earth cooled from its tumultuous formation, life found a way to emerge about 3.5 billion years ago, as suggested by the oldest sedimentary rocks that bear traces of biological activity. But what does this tell us about our existence? Scholars through the ages, from Aristotle to Darwin, have contemplated such an enigma, offering us a rich breadth of perspectives.

Recently, the discovery of *extremophiles*, organisms that thrive in extreme conditions, has expanded our understanding of how life can sustain itself in inhospitable environments. To decipher the puzzle, we must travel back to a time when Earth was a young planet whose surface was the constant scene of volcanic eruptions, lightning storms, and a primitive atmosphere filled with a cocktail of gases. Some scientists speculate that undersea hydrothermal could have provided ideal conditions for the origin of life with their constant chemical and thermal energy. Earth's early environment, often referred to as *primordial soup*, was far from hospitable by today's standards. Yet, against all odds, it was in this harsh and chaotic environment that the first precursors of life are believed to have formed. Primordial soup refers to a mixture of water, methane, ammonia, and hydrogen, subjected to the raw energy of the young Earth in the form of heat, lightning, and ultraviolet radiation. A combination reproduced in the Miller-Urey experiments showed how such conditions could generate simple organic amino acids. Here, in Earth's natural laboratory, nature was betting on itself. The occurring chemical reactions can be considered precursors to modern biochemistry, with cycles such as the *Krebs cycle* fundamental to metabolism.

From a thermodynamic perspective, such an environment was a hotbed of potential. It was a matter of increasing complexity and finding a way to store and transmit genetic information, a problem addressed by the *RNA origin paradox*. The raw materials needed for life were present, and the energy sources were abundant. The phenomenon known as *emergence* suggests that new and more complex properties can appear in complex systems. But how can one move from chaos to order? Is it just a matter of randomness? *Broken symmetry* in physics teaches us that, under certain conditions, the system can spontaneously go from a state of disorder to one of order. The second law of thermodynamics might suggest that such a mixture would forever remain a chaotic, inanimate mixture. Yet the universe, in its bold vision, had other plans in store.

Chaos theory explains how even deterministic systems can exhibit unpredictable behavior that can lead to new forms of order. Modern studies show how simple molecules began to combine in the primordial soup to form more complex compounds, according to the concept behind *supramolecular chemistry*, which studies how molecules assemble into larger structures through non-covalent interactions. One step after another, life was taking shape. In a slow, intricate, and unsurprising chemical escalation, amino acids, the building blocks of proteins, nucleotides, the components of DNA and RNA, and simple sugars, the precursors to life's energy currency, ATP, began to form. The succession of such events led to the formation of the *RNA world*, in which RNA molecules transmitted genetic information and acted as reaction catalysts. What mysterious force was driving the bubbling of such molecules? Was this perhaps the moment when matter began to *dream of* life?

The process, known as *abiogenesis* (from the Greek *a-bio-genesis*, which literally means *nonbiological origins*), was the first step in the journey from inanimate matter to living organisms and is evidenced by microscopic fossils and biochemical traces. Isotopic analysis of these traces has provided convincing evidence for the existence of primitive biological processes. In an epic journey, each step brought matter closer to the dream of autonomous life. While the formation of complex molecules decreased entropy locally, the overall entropy of the universe still increased due to the dissipation of energy in the process, in a process consistent with the principle of increasing entropy or *arrow of time* in physics. This reveals the essence of the *resilience of life*, ready to emerge against all odds and harness energy to create order out of chaos. We are the result of rebellion. Our DNA is a living archive that holds the memory of that primordial rebellion. The primordial soup was the birthplace of the precursors of life. It is within it, in a chaotic mixture of molecules and energy, that the challenge to the universe began to take shape.

The emergence of life from inorganic matter raises deep philosophical questions about the nature of life, matter, and reality. Can we see ourselves as the custodians of this mystery? It is a question that resonates with the anthropic principle, which contemplates the central role of the observer for the universe to appear as such. The transition from inorganic matter to organic life invites reflections on the essence of life, the distinction between living and nonliving, and the conditions that make the emergence of life possible. The search for extremophiles in hostile environments on Earth helps us understand the limits of such conditions. The concept of abiogenesis could also open ethical and philosophical discussions regarding the sacredness of life, its intrinsic value, and the ethical implications of understanding and manipulating the origins of life.

Despite extraordinary advances in science, transitioning from the inanimate to the living remains one of biology's most unfathomable mysteries. Research continues to expand the boundaries of possible life, as demonstrated by the recent discovery of bacteria that use arsenic instead of phosphorus, challenging our definition of life. However, scientists still need to *replicate the crucial moment when non-life becomes life*, a quantum leap that continues to elude complete human understanding. And so we still find ourselves scanning the horizon of knowledge, searching for answers to the greatest enigma.

Research has taken us from the laboratory to the vastness of space, where missions such as ESA's ExoMars look for signs of past or present life on other planets. Experiments such as those conducted by Stanley Miller and Harold Urey in 1952, which demonstrated the possibility of synthesizing amino acids from inorganic conditions similar to those on the primordial Earth, have offered fascinating perspectives but only a partial picture of abiogenesis. The discovery of complex molecular intermediates in meteorites provides clues that abiogenesis might be a cosmic process not limited to Earth alone. Modern attempts to reproduce such an event, using sophisticated prebiotic chemistry techniques, have yet to result in the creation of an autonomous living cell from inorganic materials.

NASA's *Origin of Life* project and other international initiatives are exploring the boundaries of artificial life by designing biochemical systems that mimic the functions of cells. The complexity of the origin of life is such that, despite the sequencing of genomes and the synthesis of proteins and nucleic acids, the final piece – the *spontaneous self-organization of components into a living* system – remains elusive. All this brings us back to the principle of emergence, where the sum of the parts is insufficient to explain the whole, a testament to the extraordinary complexity inherent in life and our ongoing quest to understand it. The path to the synthesis of life is littered with unanswered questions and conceptual challenges, as pointed out by researchers such as Gerald Joyce, whose work on RNA replication and molecular evolution has helped illuminate some possible avenues for understanding abiogenesis, but without unraveling the whole story. Similarly, Jack Szostak's work on prebiotic chemistry has explored how nucleic acid molecules can self-assemble into complex structures.

Scientific research is applied with dedication to the study of the sophisticated mechanisms that may have initiated the self-assembly of the first living entities on our planet, an analysis conducted through a multidisciplinary approach in which disciplines such as geology, which studies the Earth's primordial conditions, converge and integrate; biochemistry, which explores the molecular interactions essential for life; astrophysics, which questions the possible existence of life outside our solar system and seeks to understand whether the same biochemical rules can be applied universally; and synthetic biology, which through its experiments attempts to recreate or simulate biological processes to observe life emerging from nonliving conditions. Among the prominent initiatives in this field, the ENCODE (Encyclopedia of DNA Elements) project represents an ambitious undertaking that aims to map all the functional elements of the human genome. In particular, the project focuses on the vast ocean of non-coding DNA, once mistakenly thought to be genetic *junk*, which is proving to be a crucial component in understanding genetic regulation and the complex evolutionary dynamics that have led to the diversity and complexity of life as we know it. ENCODE's discoveries are helping to redefine our concept of the genome, providing new tools for interpreting the coded language of DNA and illuminating the darkness of our ignorance about the evolution of complex life once shrouded in mystery.

2.2 The role of energy in the origin of life

In its various forms, energy was abundant in the primordial terrestrial environment. A kind of *heat engine* turned (and continues to turn) heat into work. The heat from volcanic activity, electrical energy from lightning, and ultraviolet radiation from the Sun helped form the energy pool of the primordial soup. Energy was essential, for example, for the synthesis of amino acids. According to some recent hypotheses, submarine hydrothermal sources may have provided the energy needed for prebiotic chemical syntheses. We can undoubtedly say that we would not be here without that energy.

Dynamical systems theory teaches us that energy flow is fundamental to *the emergence of new states of order*. Indeed, the second law of thermodynamics states that energy always flows from high-energy areas to low-energy areas, increasing the overall entropy or disorder of the universe. Nevertheless, energy flow can drive the formation of more complex structures in the process, decreasing entropy locally. This phenomenon – known as *entropy* localization – has facilitated the emergence of life from the primordial soup. Energy-driven reactions may have been catalyzed by minerals, such as sulfides, which can act as catalysts under extreme conditions. The initial spark of life may have been lightning itself. This idea is supported by the phenomenon of *electrosynthesis*, in which electrical energy stimulates the synthesis of organic compounds. Energy from the environment was harnessed to drive chemical reactions, forming complex molecules such as amino acids, nucleotides, and simple sugars. The basic building blocks of life seem to have been forged in this way, in a crackle of energy and matter, and their formation was a crucial step in the journey from inanimate matter to living organisms. Following the vision of biochemist John E. Walker, the primordial biological building blocks accumulated and organized themselves into gradually more complex structures, like proper *molecular machinery* capable of performing specific and vital functions for the cell. Molecular machinery can be compared to engines in which chemical energy is converted into mechanical work, as is the case with ATP (adenosine triphosphate), a crucial enzyme in the process of RNA synthesis. ATP is known as the *energy currency* of living cells, underscoring the importance of such molecular systems in the cellular energy economy.

But the journey has just begun. The role of energy in the origin of life is not limited to the formation of complex molecules. The *physics of irreversible processes*, studied by Ilya Prigogine, shows us how *systems can self-organize* into progressively more complicated structures by energy flows. Energy also plays a crucial role in the formation of the first *dissipative structures*, that is, systems that maintain their structure by dissipating energy. Such systems can be compared to those studied in *non-equilibrium thermodynamics*, where small energy flows can lead to significant changes in the system's structure. Such structures, which include the first *proto-cells*, the precursors to living organisms, show how chemical reactions far from equilibrium can lead to complex, self-regulating patterns.

Could we consider life as the ultimate example of resilience? An idea encapsulated in Nassim Nicholas Taleb's concept of *antifragility*, where systems improve their ability to function in the face of shock and stress. In a sense, life can be seen as a substantial dissipative structure, a system that maintains its order by continuously dissipating energy. This is similar to the concept of a *bifurcation point*, a term from chaos theory that describes the point at which a small change can lead to a phase transition in a system. From such a perspective, the origin of life was not a singular event but a gradual process of increasing resilient complexity driven by energy dissipation.

What if energy is our deepest connection to the universe? This notion is reinforced by the *unified field theory* in physics, which seeks to describe all fundamental forces in the universe as manifestations of a *single fundamental energy field*. The idea that energy can give rise to life could evoke the concept of a divine *breath of life* or a divine creative force infusing life into inanimate matter. This parallels natural theology, where the order and complexity of the universe are seen as signs of a divine plan. In many religious traditions, energy or life force is seen as something conferred by a deity or higher force, linking the origin of life to an act of divine creation. However, from a scientific perspective, energy is simply the ability of a system to do work or produce change, as defined in the first principle of thermodynamics. Understanding how energy can transform inanimate matter into complex living structures can inspire philosophical reflections on life's essence, the universe's inherent order, and the fundamental laws that govern existence. An understanding that prompts us to explore the concepts of *dark energy* and *gravitational waves*, which make up most of the universe and may have as yet unknown implications for its origin and ultimate fate. Perhaps, by understanding the intimate nature of this energy, we can find the answers to our oldest questions. The concept of dissipative structures, where energy flows through the system to maintain order and structure, may also serve as a philosophical metaphor for exploring ideas of transformation, adaptation, and evolution. Life could be a tale of how energy incessantly transforms and evolves. The discussion of energy and the origin of life could illuminate the scientific aspects of abiogenesis and open doors toward a holistic understanding of the cosmic cycle.

2.3 The emergence of order

Order emerged from chaos, complexity arose from simplicity, and so the first steps toward life were taken. But how? From a scientific perspective, principles such as self-organization and self-assembly explain how order can emerge spontaneously in physical and chemical systems. The *principle of emergence* holds that the whole is greater than the sum of its parts, and this is evident in the complexity of life, which cannot be reduced to its constituent molecules. The energy flow through the system led to the formation of complex molecules and decreased entropy at the local level. A process was catalyzed by factors such as clay minerals and gel-disgel cycles in tide pools, which can increase the concentration of organic molecules. Random? The concept of *constructive niche* in evolutionary ecology suggests that organisms are not only passively adapted to their environments but actively contribute to their modification, creating order from disorder. A game of fate? Or an inevitable fate?

Complex molecules formed in the primordial soup, including amino acids, nucleotides, and simple sugars, began to interact with each other, driven by the flow of energy, in interactions that may have been influenced by wet-dry cycles, which could have facilitated the polymerization of nucleotides into RNA. The repetition of such interactions led to the formation of proto-cells encapsulated within a membrane, capable of harnessing energy from the environment to maintain their internal order, a crucial feature of life. The discovery of *self-replicating biopolymers* such as *ribozyme* offers a glimpse into how RNA might have played a crucial role in this process, acting both as genetic material and as a catalyst. It laid the foundation for the emergence of metabolism, the formation of the first cells, and the evolution of life. The ability to evolve is exemplified by the theory of *endosymbiosis*, which explains the origin of complex organelles such as mitochondria and chloroplasts.

The emergence of order from initial chaos is an eloquent demonstration of the inherent adaptability of life. And the role that energy plays is central. The latter drives the synthesis of ATP, the energy currency of cells, through processes such as oxidative phosphorylation. Just as an artist's touch transforms a blank canvas into a work of art, energy sculpts order from disorder. Energy not only catalyzes the formation of complex molecules but also facilitates dynamic interactions between molecules, creating a synergy that leads to the formation of more ordered and complex structures. It is a phenomenon of synergy at the heart of the science of systems and explores how interactions between the parts of a system can lead to emergent behaviors.

Marvel at the emergence of order, a process that continues to shape the evolution of life to this day. And, is this different from the very foundation of our most recent technological achievements, where the order of an artificial life could emerge from the chaos of *bits* and *qubits*? A futuristic parallel could be established with today's emerging technologies, where order and chaos are intertwined in the infinite lattice of circuits and neural networks, causing machine learning algorithms to emerge from simple rules to handle progressively more complex tasks in a process of potential *abiogenesis of artificial intelligence and life*.

Life seems to adapt and modify its conditions to resist imposed changes. Although the formation of complex structures may seem at odds with increasing entropy, it is actually an expression of how energy can be used to create order and function in an ever-changing universe. The emergence of order is a narrative of how life can find its way in a complex web of dynamic relationships, a narrative that continues to unravel as we explore the depths of life's origin and its ongoing journey of evolution. The emerging field of *systems biology* focuses on how such dynamic relationships shape biological functions, from gene regulation to cell signaling. It is a call to look beyond, to see in the fragments of knowledge the traces of a future yet to be written. As in the *holographic principle* (which we will address later), where each part contains the whole, the emergent order of life suggests that even in every cell and organism, there is a reflection of an *interconnected universe*. The processes examined appear to reflect life's deep desire to explore, adapt, and thrive, a desire that continues to drive life's evolution within a cosmic framework of change and possibility.

2.4   The first dissipative structures

As we move deeper into the primordial soup, we encounter a phenomenon that is central to the origin of life: the formation of the first dissipative structures, predicted by Belgian physicist Ilya Prigogine, who demonstrated how *complex structures capable of self-maintenance* can emerge in systems far from thermodynamic equilibrium. But what does it mean, in concrete terms, dissipative structures? Imagine a system that, despite the chaos around it, is self-regulating and self-preserving by consuming energy and restoring order. Imagine a mosaic that assembles itself, piece by piece, in the midst of a whirlwind. An amazing principle that seems to defy even the rules of imagination, but on the contrary, is the true beating heart of every living cell on Earth. Such structures, which emerged from the chaotic interaction of molecules in Earth's primordial environment, represent a significant leap in the journey from inanimate matter to living organisms. The true miracle of life consists of an orchestration of processes converging toward molecular complexity under the direction of physical laws. Is this not astounding? The process of self-organization is a fundamental aspect of dissipative structures, that is, systems that maintain their structure by dissipating energy.

The formation of dissipative structures was guided by a compact energy flow through a system. Flow is the source of order and organization, the beacon guiding matter toward self-composition and biological function. Energy from the environment was harnessed to drive chemical reactions, leading to the formation of complex molecules within proto-cells. Here, in a tiny natural laboratory, *the most extraordinary alchemy in history* took place*: the transmutation of inorganic elements into life.* And here we pause for a moment: it is, in fact, energy itself that is the creator of life. It is not only the initial spark but also the fire that shapes complexity – the unexpected genius of evolution. The first dissipative structures were a crucial step in the origin of life. The first cells were not just chemical pockets but true cathedrals of complexity, the building blocks of a tower that we continue to construct to this day. They laid the foundation for the emergence of metabolism, the process by which living organisms harness energy from the environment to maintain their structure and perform their functions. Every living thing is an intricate process that unfolds gracefully and defies simple chemical description. It is the same energy that, in renewable forms today, seeks to move the world without consuming it. And in every wind turbine and solar panel, we see a reflection of that same quest for balance, that same thirst for light that gave rise to life. They also paved the way for the formation of the first cells, the basic units of life.

Our most remote past is mirrored in the present, and in each attempt to understand and replicate such processes, we come a little closer to understanding the great enigma of existence. The evolution of dissipative structures may parallel the creation narratives where ancient gods fashion life from primordial elements. In every creation myth, in every act of shaping humans from clay, there is an intuitive recognition of the immense creative power that underlies life itself. The deity creating humans from clay, *infusing life into inanimate matter,* reminds us how each culture has tried deciphering the mystery of origin with the stories and symbols that spoke to their understanding. A very distant echo of the first dissipative structures that formed complexity from the primordial soup. It is a dialogue between past and present, between science and symbol, that continues to weave the ever more intricate and mysterious web of our beginning. In this dialogue, we are simultaneously scientists peering through the microscope and poets contemplating the stars, always searching for that *lost connection*, that bridge between the cosmos and the cell that defines man.

2.5   From the simple to the complex: the evolutionary leap

The emergence of the first dissipative structures is a witness to an extraordinary phenomenon in which inanimate matter self-organizes into stable entities capable of growth and reproduction, delineating the fundamental laws of biology from those of physics. But what triggered the paradigm shift? While not yet alive, simple dissipative structures set the stage for the evolutionary leap from simplicity to complexity that would eventually lead to the emergence of life. From simple cycles of chemical reactions to elaborate metabolic networks. Where did the first push toward order come from? The answer lies at the heart of entropy and information, where *the improbable becomes inevitable* under the direction of the laws of nature. The critical phenomenon here is self-organization, an emergent property in nonlinear systems far from equilibrium. It is as if each molecule is guided by an invisible instinct, following a matrix of order in a universe directed toward disorder. *Thermal fluctuations* and *chemical gradients* are thought to have played a key role in such a process. Such gradients are not just gradients of concentration but actual *currents of life* carving niches of order in the ocean of chaos.

The earliest dissipative structures were probably simple aggregates of molecules encapsulated in a primitive membrane. Like soap bubbles floating in a primordial wind, primitive membranes may have been the first cradles of emergent life. Recent studies suggest that the earliest membranes may have formed from *lipids* that self-assembled into *micelles* and *liposomes*. Lipids were not mere containers but dynamic tools that channeled the flow of life. They were able to harness energy from the environment to maintain internal order, a key feature of life.

They were still far from the complex, self-replicating entities we now recognize as living organisms. This represents the moment when the spark of natural selection began to ignite the engines of evolution, shaping life into increasingly intricate forms. Was it the principle of natural selection, proposed by Charles Darwin, that guided such structures toward greater complexity and efficiency? Life was sculpted by the unstoppable force of energy, a river making its way through the landscape of matter. The concept of self-organization emerging from out-of-equilibrium systems can explain how simple structures can evolve into complex systems without external design. That is, the demonstration that complexity can arise from simplicity, that order can arise from disorder, and that life, in its essence, is an *emergent phenomenon*. As energy continued to flow through these structures, it drove chemical reactions that led to the formation of increasingly complex molecules. It is a virtuous circle of energy and matter that gave rise to the wonders of life we know.

The discovery of *ribozymes*, RNA molecules that catalyze chemical reactions, provided clues on how such reactions might have organized themselves into *autocatalytic cycles*, leading to the RNA world hypothesis. The RNA world may have been the first alphabet of life, writing the instructions for complexity that continues to expand. Over time, reactions became more organized and efficient, leading to the emergence of primitive metabolic pathways.

Like the birth of a star, the process that transformed simple chemical reactions into complex metabolic systems is one of the most transcendental events in the universe's history. At the same time, the primitive membranes that encapsulated simpler structures gradually became more complex, evolving mechanisms to control the flow of molecules in and out of the structure in a phenomenon explained by the *theory of serial endosymbiosis*, which describes how the incorporation of prokaryotic organisms into the ancestors of eukaryotic cells led to greater cellular complexity. It is as if life, in the act of ingenious cooperation, decided to build more spacious and functional houses, paving the way for more evolved life forms by increasing the efficiency of energy absorption processes within the structure, enabling it to maintain its order more effectively.

The culmination of the evolutionary leap was the formation of the first cells, complex structures capable of self-replication and metabolism. Each cell became a library of information, a living archive of the history of life on Earth, and a miniature city with its traffic and pathways, energy and exchanges, a veritable microcosm of order and life. The importance of RNA as a catalyst and repository of information in the prebiotic RNA world suggests a central role in that transition. Life began to tell its story, and each RNA and each DNA are chapters in a grand epic tale. The evolution of the first cells was a milestone, with DNA later assuming the role of the primary genetic material, enabling the development of genetic codes and the transmission of hereditary information. Information transfer is the beating heart of heredity. From inanimate matter to life.

The transition from nonlife to life unfolds on time scales that defy our understanding, a journey that extends from the distant past to a future full of potential. The transition from simplicity to complexity could be understood from the narratives of ancient cultures that told how the gods forged life from primordial elements. These are potent metaphors from which to draw. Creation stories, in all their forms, are a bridge between our quest for knowledge and our need to connect with the past.

The concept of *panspermia* suggests how life may have extraterrestrial origins. This theory extends the domain of life beyond the boundaries of our planet and invites us to look to the stars in search of brothers in life. It is an idea found in the mythology of many ancient cultures. For example, the Egyptian narrative of *Khnum* modeling humans on his pottery wheel metaphorically reflects the evolutionary leap. The metaphor of modeling is universal, symbolizing the creative power that underlies every act of creation, be it a clay statue or a living cell. The myth can be paralleled by the idea of a great *cosmic genetic engineer*, a symbolic representation of the natural (or artificial?) processes that shaped life on Earth. And in this representation, we are simultaneously both sculptors and sculptures, creators of technologies that in turn shape our future.

2.6 The world of RNA

When we study the origin of life, we encounter a crucial epoch known as the *RNA world*. This biological era can be seen as nature's laboratory, where the first hypotheses of life were tested under unprecedented conditions. In the period leading up to the formation of the first cells, the conditions are created for a significant leap in the complexity of life's precursors. The RNA world hypothesis proposes that, before the advent of DNA and proteins, life was based primarily on primordial RNA molecules that acted as a store of information and as a catalyst for essential reactions, playing a pivotal role in what we might consider the dawn of life. Its ability to *self-duplicate* may have been the first step toward heredity and evolution, providing the means for experimentation and innovation in the nascent kaleidoscope of life. Dual functionality makes RNA the ideal candidate for being the first self-replicating molecule, a crucial component in transitioning from simple dissipative structures to complex living organisms. Such aggregates represented actual *molecular workshops*, where the instructions of life were transcribed and put into action.

Several lines of evidence support the RNA world hypothesis. For example, the ribosome, the site of protein synthesis in the cell, is primarily composed of RNA and not protein, suggesting the fundamental role of RNA in primordial cellular functions. Moreover, many of the most fundamental processes in modern cells, such as protein synthesis, are catalyzed by RNA molecules. Experiments have shown that RNA molecules can self-replicate under certain conditions, giving sufficient scientific credence to the idea that RNA may have been the first genetic material.

The existence of the RNA world challenges us to rethink evolution as a process not only of selection but also of pure molecular innovation. The transition from the RNA world to the DNA-protein world we know today would have been driven by the same forces that led to the formation of the first dissipative structures – the flow of energy and the laws of thermodynamics. DNA would have brought stability to the chaos of chemical reactions, while proteins would have introduced a diversity of functions, thus opening the door to new evolutionary possibilities. DNA, with its more stable double-helix structure, would gradually take over the role of genetic repository. At the same time, with their greater catalytic versatility, proteins would assume the role of metabolic catalysts.

RNA's inherent potential to serve as both an information store and a catalyst for reactions could be interpreted as a manifestation of the sophisticated simplicity ingrained in the laws of the universe, where everything has a place and a purpose. The world of RNA conjures up images of a primordial epoch, a stage of metamorphosis where the functional duality of RNA, a repository of information and catalyst of reactions, is a concrete manifestation of the interconnectedness of inorganic matter and the complexity of life. The world of RNA is not just a past phase of evolution but a bridge to a deeper understanding of life itself.

2.7 The advent of DNA

The exponential increase in the complexity of life made it necessary for a more stable and efficient molecule to carry genetic information. This is where DNA comes in. The switch from RNA to DNA is a fundamental paradigm shift in the history of life, marking the transition from a provisional inheritance system to a more permanent and robust one. DNA, with its double-helix structure, is more stable than RNA and less prone to mutation. A transition that has allowed the conservation of longer genetic sequences and the possibility of developing DNA repair mechanisms, increasing the fidelity of inheritance. Its stability makes DNA an ideal molecule for storing genetic information, especially when life forms are complex and the amount of genetic information is very large.

The storage of genetic information in DNA thus paved the way for the explosion of biological diversity, with each DNA sequence becoming a potential for a new life form. The transition from RNA to DNA was a gradual process, driven by the forces of evolution. Like the slow crystallization of minerals from a saturated solution, natural selection refined living systems toward more stable and enduring genetic structures. With their greater stability and efficiency, DNA-based life forms would have had an evolutionary advantage over RNA-based life forms, the true beginning of an evolutionary race that led to the complexity we observe in the living realm today.

Over time, DNA would assume the role of genetic storage, while RNA would be relegated to the role of messenger, carrying genetic instructions from DNA to the cell's protein machinery. Thus, RNA went from protagonist to supporting, but still indispensable in the flow of genetic information. It marked the transition from the RNA world to the DNA-protein world we know today, setting the stage for the emergence of metabolism and the formation of the first cells. An evolutionary event that allowed life to become more deeply rooted in the material world, stabilizing life forms in an ever-changing landscape. The advent of DNA could be interpreted as a manifestation of the intricate order and predictability needed to sustain life in a more evolved and complex form.

DNA has become the core of life, the central repository in which the codes for *biodiversity* are inscribed, a repository of potential allowing species to adapt and thrive in every corner of the planet. Evolution from RNA to DNA can be likened to a kind of second creation or a refinement stage in the creation myth. An intervention that is as natural as it is divine or alien.

In many mythological traditions, there are stories of a primordial creation followed by a period of refinement or consolidation, where what is created becomes more stable and less subject to mutation, representing the maturation of life from its youthful and stormy state to an adult phase of greater security and reliability. In DNA, it is possible to see the powerful symbol of life's ingenuity. This molecular architecture tells us as much about our history as it does about our future's infinite possibilities.

2.8 The birth of metabolism

Continuing our exploration of the origin of life, we come to a new pivotal moment: the emergence of *metabolism*. Metabolism is an intricate labyrinth of biochemical reactions that represents the alchemy of life. It can transform inorganic elements into complex webs of life and enable living organisms to extract energy from the environment and use it to maintain their structure and perform their functions. This is a fundamental characteristic of life.

In the early stages of life's evolution, the first dissipative structures, encapsulated within primitive membranes, could harness energy from the environment to maintain internal order. The first cells were like biochemical jewels. Using light from the sun, heat from the earth, or minerals from the sea to power the earliest life forms evolved into more complex entities, into a more efficient and organized system for collecting, preserving, and using energy. This is where metabolism comes in. Metabolism began as a relatively simple mechanism but soon evolved into a system of interconnected reactions.

The emergence of metabolism was a gradual process, driven by the forces of evolution. The earliest metabolic pathways were probably simple chemical reactions catalyzed by primitive RNA molecules or proteins that allowed proto-cells to harness energy from the environment. Such primitive reactions were the foundations on which the complex edifices of life would be built. Over time, the pathways became increasingly complex and efficient, leading to the diverse range of metabolic processes we see in modern cells.

The advent of DNA, with its stable double-helix structure, played a crucial role in the evolution of metabolism. DNA acted as the keeper of *metabolic recipes*, ensuring that the banquet of life could continue through the generations. DNA provides a reliable way to store and transmit instructions for metabolic processes, allowing them to be passed on from generation to generation. This allowed metabolic pathways to become more complex and efficient over time because while beneficial mutations were conserved, harmful ones were eliminated. Metabolism, in its complexity and precision, highlights a form of intelligence or design that seems to defy randomness, inviting reflection on the purpose that might exist behind the mechanisms of life.

2.9 The first cells

The first cells probably evolved from the proto-cells that populated the RNA world, ancient molecular aggregates that laid the foundation for life as we know it. Primitive structures comprise a simple membrane enclosing RNA molecules and small metabolic networks. They were the first to display some of the fundamental features of life. The transition from rudimentary molecular architecture to a full-fledged cell is one of the great puzzles that science is still trying to decipher. Structures are still far from the complex, self-replicating entities we now recognize as cells. The transition from proto-cells to actual cells can be compared to one of the most significant evolutionary leaps. As energy continued to flow through these structures, it drove the formation of increasingly complex molecules and metabolic networks. The first cells marked a fundamental change in the game of life, introducing the concept of heredity and biological evolution. With the advent of the first cells, life began experimenting with ever-new forms, functions, and strategies, like an artist discovering new colors and brushes. They were capable of self-replication, could harness energy from the environment to maintain their structure and perform their functions, and contained a genetic system that allowed them to evolve over time.

Religious traditions that see a divine design in creation could interpret the evolution of the first cells as a significant step in the divine plan, reflecting a process of order and purpose that reflects the intelligence or intention behind the growing complexity of life. An exploration leads us to a deeper understanding of ourselves, as every cell in our bodies is descended from that primordial birth event. It may also raise questions about identity, change, and the nature of evolution, offering fertile ground for reflections on the dynamics of being and becoming. The birth of the first cells marks a pivotal milestone in life's journey, a moment of radical transformation that led to the diversity and complexity of life we observe today, a reminder of our intimate and indestructible connection to all living things. In a sense, it symbolizes the promise and potential inherent in life itself, its innate tendency toward complexity, adaptation, and evolution.

The struggle against entropy also drove the evolution of primitive life. Mutations, essentially random increases in genetic entropy, led to new traits. Each mutation was nature's gamble, a genetic roll of the dice with the potential to unlock new capabilities. If the characteristics improved the cell's ability to harness energy and maintain its structure, they were preserved and passed on to future generations. Natural selection was thus the great judge in this struggle, rewarding solutions that increased the robustness and flexibility of early life forms. Over time, this process led to the evolution of more complex and efficient life forms. And so, from the simple cell to complex ecosystems, life has continued to rewrite entropy rules, creating a diversity of forms and functions that defy imagination.

From the Big Bang to Consciousness

Notes and insights to Chapter 2

1. Exploration of the Earth's environment during the early days of Earth's formation. See: Knoll, A.H. (2003). *Life on a Young Planet.*

2. Explanation of the process of abiogenesis and the formation of complex molecules. For more details, see: Hazen, R.M. (2005). *Genesis: The Scientific Quest for Life's Origins.* Discussion of the formation of amino acids, nucleotides and simple sugars in the primordial soup. For more details, see: Sutherland, J.D. (2016). *The Origin of Life - Out of the Blue.*

3. Philosophical reflection on abiogenesis and the ethical and philosophical implications of understanding the origins of life: Bedau, M.A., and Cleland, C.E. (2010). *The Nature of Life.*

4. Miller, S. L., & Urey, H. C. (1959). *Organic Compound Synthesis on the Primitive Earth.* Science, 130(3370), 245-251. It is a classic study that provided the first experimental evidence of how organic components essential for life can be synthesized from inorganic compounds under conditions that mimic those on the primitive Earth. Joyce, G. F. (2002). *The antiquity of RNA-based evolution.* Nature, 418(6894), 214-221. Gerald Joyce is a leading researcher in the field of the origin of life, and his work has explored the possibility that RNA may have played a crucial role in abiogenesis. Szostak, J. W. (2012). *Attempts to define life do not help to understand the origin of life.* Journal of Biomolecular Structure and Dynamics, 29(4), 599-600. Jack Szostak is another researcher who has investigated the origin of life, particularly the role of primitive cell membranes and self-replication. The cited sources represent only a fraction of the extensive and continuing research in abiogenesis and the origin of life.

5. Discussion of philosophical and religious interpretations of energy and its role in the origin of life: Clayton, P. (2004). *Mind & Emergence:*

From Quantum to Consciousness. Reflections on the Essence of Life: Philosophical reflections on the essence of life, the inherent order in the universe, and the fundamental laws governing existence. For further details, see: Davies, P. (1999). *The Fifth Miracle: The Search for the Origin of Life.*

6. Analysis of early cell formation as a culmination point in the evolutionary leap. For further details, see: Schopf, J.W. (1993). *Microfossils of the Early Archean Apex Chert: New Evidence of the Antiquity of Life.*

7. Detailed discussion of the RNA world hypothesis and its role in the origin of life: Gilbert, W. (1986). *The RNA World.* Analysis of the transition from the RNA world to the DNA-protein world. For further details, see: Bernhardt, H.S. (2012). *The RNA world hypothesis: the worst theory of the early evolution of life (except for all the others)*; Orgel, L. E. (2004). *Prebiotic chemistry and the origin of the RNA world.* Critical Reviews in Biochemistry and Molecular Biology, 39(2), 99-123.

8. Reflection on the functional duality of RNA as an information store and reaction catalyst. For more details, see: Joyce, G.F. (2002). *The antiquity of RNA-based evolution.*

9. Exploring the transition from RNA to DNA in the context of the evolution of life. For further details, see: Forterre, P. (2002). *The origin of DNA genomes and DNA replication proteins.* Discussion of the greater stability of DNA compared to RNA. For further details, see: Kornberg, A., & Baker, T.A. (1992). *DNA Replication.*

10. Exploring the evolution of metabolism from early dissipative structures to more complex entities. For further details, see: Martin, W., & Russell, M.J. (2003). *On the origins of cells: a hypothesis for the evolutionary transitions from abiotic geochemistry to chemoautotrophic prokaryotes, and from prokaryotes to nucleated cells*; Wächtershäuser, G. (1988). *Before enzymes and templates: theory of surface metabolism.* Microbiological Reviews, 52(4), 452-484.

11. Discussion of the role of DNA in the evolution of metabolism and the transmission of metabolic information. For more details, see: Lane, N., & Martin, W. (2010). *The energetics and genetics of the first cells.*

12. Analysis of the transition from proto-cells to cells and the factors that drive it. For further details, see the Nobel Prize-winning text: De Duve, C. (1991). *Blueprint for a Cell: The Nature and Origin of Life.*

Neil Patterson Publishers.Lazcano, A., & Forterre, P. (1999); *The molecular search for the last common ancestor*. For further discussion of the evolution of the first cells and their complexity, it is advisable to examine the work of Woese, C. (1998). *The universal ancestor*. Proceedings of the National Academy of Sciences, 95(12), 6854-6859; Margulis, L. (1970). *Origin of eukaryotic cells*. Yale University Press.

13. Reflection on the implications of the emergence of the first cells for the understanding of life and reality. For further details, see: Maturana, H., & Varela, F. (1980). *Autopoiesis and cognition: The realization of the living*.

From the Big Bang to Consciousness

Chapter 3

THE LANGUAGE OF LIFE: DNA

From the Big Bang to Consciousness

From the Big Bang to Consciousness

The capacity to blunder slightly
is the real marvel of DNA.
Without this special attribute,
we would still be anaerobic bacteria
and there would be no music.

LEWIS THOMAS

A Long Line of Cells, Collected Essays, 1990

From the Big Bang to Consciousness

3.1 The structure of DNA

Before analyzing the complex relationship between DNA and entropy, it is essential first to understand the structure of DNA itself. *Deoxyribonucleic acid,* better known as DNA, is a molecule that encodes the genetic instructions of all known living organisms and many viruses. Discovered in 1953 by James Watson and Francis Crick, DNA has revolutionized the field of molecular biology. Its primary function is to *store the information* that guides the development and functioning of living organisms, making it one of life's most crucial inventions.

DNA is a double-stranded molecule, with each strand consisting of a long sequence of four different types of molecules or bases: *adenine* (A), *guanine* (G), *cytosine* (C), and *thymine* (T). The bases are organized into specific sequences, which can be likened to letters in a very long instruction book. They are attached to a sugar-phosphate backbone, forming what is known as a *nucleotide.* Nucleotides link together to form the long strands of DNA. Its structure can be changed by factors such as ultraviolet radiation and chemicals, which can cause mutations. In the double-helix structure, strands wrap around each other, with bases pairing in the middle: adenine always pairs with thymine, and cytosine always pairs with guanine. Base complementarity is crucial for DNA replication accuracy and the genetic code during protein synthesis. Base pairings are critical for DNA replication, a process essential for cell division and the propagation of life. When a cell divides, the two strands of the DNA molecule unwind and separate. The process is catalyzed by enzymes such as helicase and DNA-polymerase. Each strand then serves as a template for creating a new complementary strand, resulting in *two identical copies* of the original DNA molecule. *Semi-conservative replication* is essential for preserving inherited characteristics and ensuring the faithful transmission of genetic information from one generation to the next. Nonetheless, occasional copying errors or *mutations* can lead to *genetic variations* that drive evolution.

From a thermodynamic perspective, the structure of DNA represents a delicate balance. The stability of the double helix is maintained by hydrogen bonds between base pairs and hydrophobic interactions that keep the bases inside, away from surrounding water molecules and disturbed during processes such as DNA replication and RNA transcription. DNA replication, on the other hand, requires strand separation, which involves overcoming the energy barrier provided by such stabilizing forces. It exemplifies how life overcomes thermodynamic barriers to maintain and transmit life itself. The balance between stability and accessibility is essential for DNA to function as a carrier of life's genetic information. In addition, DNA can undergo repairs through error-correcting mechanisms, demonstrating an amazing capacity for *self-preservation*.

Thus, at its core, the structure of DNA provides a stable yet dynamic storehouse of information that can withstand the onslaught of entropy while facilitating the creation of life's extraordinary diversity and complexity. Future research may yet reveal much about how this incredible molecule continues to evolve and adapt in a constantly changing universe.

3.2 DNA: A thermodynamic miracle?

How does a molecule like DNA, ordered and complex, emerge and persist in a universe governed by entropy? From a philosophical point of view, the very existence of DNA testifies to the inherent complexity of the universe, which is also expressed through self-organizing mechanisms that contradict the growing disorder. In fact, the second law of thermodynamics does not conflict with the complexity of life; instead, the law provides the context in which life can emerge and develop.

DNA complexity can be interpreted through the concept of *informational entropy*, which differs from physical entropy. While the latter is concerned with molecular disorder, informational entropy deals with the loss of information. DNA, by maintaining information through repair and replication mechanisms, shows how order can be maintained and even amplified in living systems despite physical entropy. Although DNA is undoubtedly a structure of remarkable order, its formation and maintenance are processes based on thermodynamics. The creation of the DNA double helix involves joining individual nucleotides into a stable structure, a process that accounts for a decrease in entropy. The formation of the DNA double helix is driven by the energy released when nucleotides come together. The release of energy is facilitated by the hydrogen bonds formed between the bases and the hydrophobic interactions that keep the bases stacked together at the center of the helix, away from water. In the context of the surrounding environment, the process of DNA formation leads to a net increase in entropy, and the role of water as a biological solvent is critical. The structure of water allows the formation of a network of hydrogen bonds that is as robust as it is dynamic, thus facilitating the molecular interactions necessary for DNA formation and replication.

Thermodynamics also supports DNA replication, a process that creates order out of chaos. The replication process begins with separating the DNA double helix, an action that requires an energy input to break the hydrogen bonds between the bases. The energy is supplied by a series of specialized proteins and is procured by the cell's metabolism. Once the strands are separated, new nucleotides are recruited to form complementary strands, guided by the base pairing rules. The formation of new filaments is energetically favorable and leads to a release of energy, advancing the process. This phenomenon is an example of how biological systems use Gibbs free energy to carry out processes that, while involving a local decrease in entropy, are in agreement with universal physical laws. The concept of Gibbs free energy links entropy and enthalpy into a single value that predicts the spontaneity of a reaction.

The structure of DNA and the processes it facilitates – the storage, replication, and transmission of genetic information – represent another way life ingeniously navigates our universe's thermodynamic flows. From a broader perspective, DNA can be seen as a mechanism through which the universe explores its own complexity.

3.3 The role of DNA in energy dissipation

Although the primary function of DNA is to store and transmit genetic information, it is also essential to consider it from the perspective of energy and entropy. DNA plays a crucial role in the overall energy economy of the cell and, by extension, in the entropy management strategy of life. It is not only the stability of DNA that is relevant but also its dynamics during the cell cycle. Indeed, cells expend a considerable amount of energy in DNA replication and repair, processes that are critical for survival and evolution. Their energy expenditure is an example of how biological systems are not isolated systems but open systems, exchanging energy and matter with the environment and increasing external entropy in a way consistent with the second law of thermodynamics. In the context of thermodynamics, the formation of the DNA double helix can be seen as an energetically favorable process. As we mentioned earlier, the combination of nucleotides into a stable structure releases energy, mainly through the formation of hydrogen bonds and hydrophobic interactions. Notably, the release of energy during the formation of the DNA double helix is an example of a spontaneous process that increases the universe's entropy in a non-intuitive way.

Once formed, the DNA molecule provides a stable and energy-efficient means of storing genetic information. Its double-helix structure protects genetic information from degradation, which releases energy and increases entropy. By maintaining the integrity of information over time, DNA allows the continuity of life processes that actively dissipate energy and increase the overall entropy of the universe.

DNA's function as an information store parallels its function as a catalyst for processes that increase entropy, such as cellular respiration and biosynthesis. In such processes, energy stored in organic molecules is converted into usable forms for cellular work, while entropy is expelled in the form of heat and molecular waste. DNA replication also has important thermodynamic implications. It is an energy-consuming process, offset by the energy released during the formation of new DNA strands. Thus, DNA replication contributes to the flow of energy through the cell, resulting in an increase in entropy. The DNA replication cycle is an example of how living systems use carefully regulated mechanisms to ensure that energy is used efficiently and entropy is managed in a way that sustains life. In addition, the genetic information encoded by DNA directs the synthesis of proteins, which are the cell's workhorses and perform a myriad of functions in it, including catalyzing chemical reactions, transporting molecules, and carrying out the cellular processes that enable life to persist and reproduce.

The complex intertwining of DNA and the proteins it synthesizes highlights the *cyclic nature of energy and information* in life forms, underscoring DNA's role not only as a *genetic library* but also as the *dynamic engine* behind the complexity of life. Therefore, DNA, in its *in-formative* structure and function, is the *repository of the blueprint of life*, hinging the dynamic balance between order and chaos, stability and change, accumulation and dissipation of energy. The role of DNA in managing the flow of energy and matter in life forms is further evidence of life's ingenious challenge to the apparent fate of the universe.

3.4 The evolution of the genetic code

At first glance, the language of DNA – the genetic code – seems simple. Four different bases – *adenine, guanine, cytosine,* and *thymine* – are arranged in various sequences to create genes, which in turn code for proteins. Yet behind its apparent simplicity lies an incredibly intricate system, the result of evolution over billions of years aimed at creating and sustaining life's vast diversity and complexity. Like human languages that have evolved to express increasingly abstract and complex concepts, the genetic code has undergone a parallel evolution, developing the ability to encode increasingly complex biological information. In its earliest form, the genetic code probably consisted of shorter sequences and fewer amino acids.

Over time, the genetic code has expanded and diversified, driven by mutation and natural selection. Interestingly, this process of expansion and diversification mirrors the principles of *cultural selection*, where human ideas and inventions evolve and specialize in response to environmental and social pressures. Diversification has increased the complexity of life forms, facilitating their adaptation to different environments and lifestyles.

In addition, the evolution of the genetic code has enabled more efficient use of available resources, including energy sources. For example, the emergence of more complex proteins has enabled the development of metabolic pathways capable of extracting energy from different types of food sources. Bioenergy efficiency is thus analogous to the human search for the most sustainable and efficient energy systems. Pathways are traced through a series of chemical reactions that transform energy and matter, contributing to the overall increase in entropy.

The genetic code has also evolved *redundancy*, known as *degeneracy*, in which multiple codons (sequences of three DNA bases) encode the same amino acid. A *functional overabundance* comparable to the presence of synonyms in the same language that can enrich language while maintaining precision of meaning. A feature that adds a level of *error tolerance* to the genetic code, allowing genetic mutations that can lead to variation and evolution without necessarily disrupting protein synthesis.

Through the genetic code, life evolves elegantly on the edge of thermodynamic equilibrium, encoding the blueprints of increasingly complex life forms while enhancing the capacity for innovation, adaptation, and survival.

3.5 Genetic variation

The role of DNA is not limited to faithfully storing and transmitting the genetic code. An integral part of its function is *generating variation*, the evolution's raw material. Genetic variation, resulting from processes such as *mutation* and *recombination*, introduces differences in DNA sequences among individuals in a population. Differences that can affect the characteristics of organisms, affecting their ability to survive, reproduce and adapt to the environment. The ability to generate variability is critical to biological resilience and flexibility, allowing species to thrive in a wide range of environmental conditions.

Viewed through the lens of thermodynamics, genetic variation presents another dimension of life's strategy to manage entropy. Genetic variation increases the diversity and complexity of life, which, paradoxically, may seem contrary to entropy's tendency toward disorder. However, the apparent contradiction is resolved when we consider that increased complexity within the biosphere is offset by increased disorder, or entropy, in the surrounding environment. Life creates internal order and promotes increased entropy in its environment through *metabolic* and *ecological* processes.

Consider genetic mutation, a significant source of genetic variation. Mutations occur due to various factors, such as errors during DNA replication or exposure to certain chemicals or radiation. Although mutations introduce changes in DNA sequence, they also require energy to occur. For example, the process of DNA repair (i.e., a response to damage that causes mutations) requires a lot of energy, which, as we mentioned, turns out to be essential for evolutionary innovation. Thus, while introducing variation and complexity, the mutation process also contributes to energy dissipation and a net increase in entropy.

Recombination, another source of genetic variation, involves the exchange of DNA segments between homologous chromosomes during sexual reproduction. This process creates new combinations of genes, contributing to the diversity of traits within a population. Recombination is also energy-consuming, contributing to the overall flow of energy and increasing entropy. In addition to its biological function, recombination is a brilliant example of how biological systems can exploit random mechanisms to generate order and function.

Genetic variation resulting from mutation and recombination provides the raw material upon which natural selection acts, leading to the evolution of life's immense diversity and complexity. A dynamic process of variation and selection that, while creating localized order, ensures a continuous dissipation of energy, ultimately driving the increase of entropy in the universe and reminding us that evolution is not a process that aims for static perfection but rather a dynamic of constant change, allowing life to adapt and thrive amid thermodynamic chaos. Thus, genetic variation, an integral part of DNA function, is part of the strategies that life adopts to navigate the laws of thermodynamics. It exemplifies how life simultaneously creates complexity and fosters energy flow, ingeniously defying the march toward universal disorder.

3.6 The role of sexual reproduction

Sexual reproduction is a defining feature of most eukaryotic life forms, *including humans*. It involves combining the genetic material of two parents to create genetically diverse offspring from both, a process fundamentally intertwined with the role of DNA and, from a thermodynamic perspective, a powerful tool in life's strategy for managing entropy. Sexual reproduction enables genetic material recombination, increasing life's diversity and complexity. Consistent with life's challenge strategy, the increased complexity is balanced by the energy required for the processes involved in sexual reproduction and the resulting increase in entropy in the surrounding environment. The consequent increase in complexity, in fact, is an example of *emergent order*, a phenomenon that occurs in spite of – or perhaps because of – thermodynamic laws governing physical systems.

Consider *meiosis*, the specialized form of cell division involved in sexual reproduction. It is an energy-intensive process, as it involves DNA replication, intricate chromosome movements, and *cytokinesis*. In addition, the recombination events during meiosis involve breaking and rejoining DNA strands, processes that also require energy. Thus, although meiosis contributes to increased genetic variation and complexity, it also promotes energy flow, contributing to increased entropy in the universe. It is important to recognize that such processes increase genetic variability and incorporate mechanisms for repair and error control, demonstrating life's ability to evolve in ways that protect the integrity of genetic information while exploring new evolutionary possibilities.

Sexual reproduction also has indirect thermodynamic implications. The genetic diversity it generates provides the raw material for evolution, enabling populations to adapt to their environments. Adaptations that often involve changes in the way organisms interact with and extract from the environment. For example, adaptations may allow an organism to use a new food source, leading to new pathways for energy flow and increased entropy. Adaptations can also affect *biogeochemical cycles* and ecosystems globally, influencing the flow of energy and matter across the planet and among different life forms.

Moreover, many of the traits that evolve through sexual reproduction, such as elaborate displays of courtship or competitive behaviors, require significant energy investment. Behaviors that – while increasing the chances of reproductive success – also ensure continuous energy dissipation, aligning with the thermodynamic mandate to increase entropy. Sexual selection – which often rewards traits that require greater energy investment – underscores how evolutionary drives can work in synergy with thermodynamic principles. Sexual reproduction, by promoting genetic diversity and driving evolutionary adaptation, thus plays a critical role in the thermodynamic balancing act of life. It emphasizes how life, through DNA, defies the seeming inevitability of universal disorder, creating localized pockets of complexity while ensuring the general increase in entropy, a dynamic that is a reflection of life's amazing ability to generate new forms and functions through constant dialogue with universal physical forces.

3.7 Epigenetics: another level of complexity

Navigating the intricate interplay between life and the laws of thermodynamics, we encounter epigenetics – the study of changes in organisms caused by altering gene expression rather than the genetic code itself. Epigenetic modifications, including DNA *methylation* and *histone modification*, alter how cells read genes and, consequently, how they function, contributing to the complexity of life.

Epigenetic changes can have profound effects on an organism. They can determine whether a gene is *on* or *off*, affect the amount of expression of a gene, and even how a gene *responds* to environmental signals in a dynamic that represents the mechanism by which organisms can respond and adapt rapidly to environmental changes without changes to the underlying genetic information. Epigenetics can be considered a *cellular memory system* that records environmental events and regulates gene response. Epigenetics offers an additional level of control over genetic information, allowing organisms to fine-tune the function of genes according to the environmental context, an extra level of regulation comparable to a conductor's ability to *modulate* the intensity and timbre of a musical work without changing the notes written on the score.

From a thermodynamic perspective, epigenetics adds another dimension to life's strategy for managing entropy. Epigenetic modifications, like other biological processes, require energy. They involve the addition or removal of chemical groups to or from the DNA molecule or the histone proteins around which DNA is wrapped. The process underscores the complex interplay between energy and information in biological systems, where genetic information is stored and modulated in a framework of continuous energy change. In addition, epigenetic changes can lead to increased complexity. They allow cells with the same genetic material to differentiate into various cell types, each with a unique set of active and inactive genes. A fundamental phenomenon in determining the functional specializations of cells, such as differentiation between cells of the nervous system and those of the muscular system. Once again, the increase in order within the organism is balanced by the resulting increase in entropy in the surrounding environment.

Environmental factors can also influence epigenetic changes and can sometimes be transmitted from one generation to the next. This suggests that *an organism's experience can affect not only its own phenotype but potentially that of its descendants*, a concept that challenges traditional conceptions of heredity and natural selection. This means organisms can inherit adaptations to environmental conditions experienced by their parents or even earlier ancestors, providing another pathway for evolution and adaptation to changing environments.

Epigenetics adds another layer of complexity to DNA's role in entropy management. It reflects the extraordinary ability of organisms to encode, store, and transmit information through mechanisms that transcend simple nucleotide sequencing.

Epigenetics exemplifies how life, through multiple levels of control over genetic information, can enhance its adaptability and survival while ensuring continuous energy dissipation and increasing entropy.

3.8 DNA damage and repair

As the keeper of the *code of life,* DNA is not invulnerable. It continually faces threats that can damage its structure and disrupt its informational function, including environmental factors such as radiation and chemicals and normal cellular processes such as DNA replication and metabolism. Fortunately, life has developed sophisticated DNA repair mechanisms to correct damage and maintain genetic integrity. The existence of DNA damage and repair mechanisms offers another demonstration of life's strategy to manage entropy. The phenomenon can be observed as an exemplification of the dichotomy between equilibrium and disequilibrium, in which biological systems move away from thermodynamic equilibrium to maintain internal order in the face of external disorder. DNA damage introduces disorder into the genetic material, aligning with the universal trend toward increasing entropy. However, incorrect damage to DNA would result in mutations that could potentially disrupt essential biological processes, threatening the organism's survival and reproduction.

To counter this phenomenon, life has perfected a variety of DNA repair mechanisms, including *base excision repair, nucleotide excision repair, mismatch repair,* and *double-strand break repair.* The diversity of repair mechanisms reflects the complexity of possible DNA injuries and the vital importance of maintaining genetic integrity in a changing environment. The processes involve damage detection, removal of the damaged part, and synthesis of a corrected sequence using the undamaged strand as a template. DNA repair mechanisms require energy, including ATP, the *biological currency* of energy, and energy investment, which is a manifestation of the law of conservation of energy applied to biological systems: energy is neither created nor destroyed but transformed to maintain genomic stability. Enzymes involved in DNA repair processes, such as *helicases*, *nucleases*, *polymerases*, and *ligases*, use ATP to carry out their functions. Therefore, while DNA repair restores order in the genetic material, it ensures a continuous flow of energy, which drives the increase of entropy in the universe.

The energy expenditure associated with DNA repair is significant. Constant surveillance and repair of DNA are investments that reflect the law of *opportunity cost*, where resources expended on repair are diverted from other cellular functions. Studies have shown that cells can devote significant resources to DNA repair, especially in response to extensive damage. Such investment underscores the critical importance of maintaining genetic integrity for the survival and reproduction of life.

DNA repair also has implications for genetic variation and evolution. Mutations introduced accidentally during repair can be sources of genetic variability that, if advantageous, can be selected for and fixed in the population. Errors during DNA repair can introduce mutations, contributing to genetic variation. In addition, some DNA repair mechanisms, such as nonhomologous end splicing involved in the repair of double-strand breaks, are error-prone, potentially driving evolutionary change.

The existence and functioning of DNA repair mechanisms illustrate the delicate balance that life achieves to manage entropy. Each DNA repair action can be viewed as a battle won against increasing disorder, a symbol of life's resistance to entropy. As DNA damage aligns with the trend of growing disorder, life invests considerable energy in repairing it, restoring order to the genetic material, and ensuring the continuity of life processes.

3.9 The human genome

The human genome, composed of some 3.2 billion DNA base pairs, is an extraordinary testament to life's challenge to the second law of thermodynamics. A marvel of complexity, it contains the blueprints of the human body, instructing the creation of everything from the smallest proteins to the most intricate neural networks. The genome can be seen as a vast library, where each book is a gene with its precise location and function. Organizing billions of nucleotides into precise sequences to encode biological information is an incredibly orderly state of matter. Its structure results from a long evolutionary history, during which organisms have had to survive and draw energy from the environment, transforming it into a valuable form for building and maintaining their internal complexity.

The creation and maintenance of the human genome involves an energy expenditure that contributes to the overall increase in entropy in the universe. Expenditure that occurs during DNA replication, repair, and transcription. For example, DNA replication – which arises every time a cell divides – involves unwinding the DNA double helix and synthesizing new strands complementary to the original ones, a process that requires considerable energy, mainly in the form of ATP, to power the enzymes involved that function as *molecular machines*, with precision mechanisms that are the product of millions of years of evolutionary optimization. In particular, the synthesis of each new DNA nucleotide helps to increase entropy by balancing the order created by the precise copying of genetic information.

The complexity of the human genome also enables a range of functions. In addition to protein coding, it also includes regions involved in regulating gene expression, determining chromosome structure, and safeguarding genome stability, among other roles. *Non-coding* regions, once considered *junk DNA*, have proven to be essential for the overall regulation and functionality of the genome. They are functionalities that enable humans to adapt to a wide variety of environments and situations, enhancing survival and reproduction.

The human genome has the potential for immense future complexity. New genetic variants can be created through processes such as mutation, recombination, and natural selection, some of which could drive evolution and increase human adaptability. A process of genetic variability that is a source of biological diversity but also an example of how genetic information can be reformulated in response to new environmental challenges and opportunities. It is testimony to how life, in its continuous development, represents a *dynamic creative force* within the laws of physics.

3.10 The future of DNA: genetic engineering

Looking to the future, one cannot help but reflect on the role of DNA in the evolution of life and the implications of our increasing ability to manipulate it. Advances in genetic engineering, particularly techniques such as CRISPR-Cas9, have given us unprecedented power to modify the genetic code. Such technologies not only enable the elimination of genetic diseases but also open the door to possibilities such as the creation of organisms with tailored traits or even the resurrection of extinct species.

The ability to influence the course of life's evolution presents an interesting thermodynamic perspective. The act of rewriting DNA is an expression of our growing understanding and control over biological information, which is the beating heart of living things. Genetic engineering, like natural processes involving DNA, requires energy expenditure. Designing and executing a genetic editing protocol requires both human intellectual energy and physical resources. The apparent paradox reflects the complexity of interpreting the laws of thermodynamics in highly organized biological systems. The more precise and complex the genetic alteration, the greater the input of energy required and, thus, the greater the entropy produced. Moreover, by enabling more targeted and precise modifications to the genome, genetic engineering could accelerate the creation of biological complexity, opening up new evolutionary pathways that would otherwise have been foreclosed, making evolution no longer a process driven solely by chance and natural selection but also by human intention. This could lead to new life forms or an increase in existing ones, both representing local entropy decreases offset by the associated energy expenditure.

The emergence of genetic engineering technologies can be seen as an extension of life's thermodynamic strategy: creating order and complexity by contributing to the increase in the universe's overall entropy. It is another demonstration of how life uses energy to organize and adapt in increasingly sophisticated ways.

The ethical implications of genetic engineering are immense and beyond the scope discussed here. However, it is noteworthy to consider them in the thermodynamic context. The ethical debate revolves around the question of how and whether we should use our power of genetic manipulation, taking into account respect for the autonomy of nature and the dignity of life in all its forms. The potential to create life forms with greater complexity and adaptability could help life in its perpetual challenge to entropy, and as our power over life designs increases, so does our responsibility to exercise such power wisely, with foresight and respect for the complex interplay between life and the laws of the universe.

The future of DNA, influenced by our ability to manipulate it, is a frontier that confronts us with choices that could define the course of biological evolution and the conditions of life on Earth for generations to come. As life's genetic code continues to evolve, naturally or through human intervention, the fight against entropy continues, reinforcing life's position as the universe's challenge against disorder.

From the Big Bang to Consciousness

Notes and insights to Chapter 3

1. For an understanding of DNA structure, refer to: Watson, J. D., & Crick, F. H. (1953). *Molecular Structure of Nucleic Acids: A Structure for Deoxyribose Nucleic Acid.* Nature, 171(4356), 737-738; Alberts, B., Johnson, A., Lewis, J., Raff, M., Roberts, K., & Walter, P. (2002). *Molecular Biology of the Cell.* Garland Science.

2. References: Styer, D. F. (2000). *Entropy and evolution.* American Journal of Physics, 68(12), 1090-1093.

3. DNA is not only a repository of genetic information, but also plays a crucial role in the energy economy of the cell, managing the entropy strategy of life. References: Morange, M. (2008). *The origins of life: understanding the process of abiogenesis.* Advances in Physics: X, 1-20.

4. For insights into the evolution of the genetic code: Maynard Smith, J., & Szathmáry, E. (1995). *The Major Transitions in Evolution.* Oxford University Press; Crick, F. H. (1968). *The origin of the genetic code.* Journal of Molecular Biology, 38(3), 367-379.

5. For insights into genetic variation: Hartl, D. L., & Clark, A. G. (2007). *Principles of Population Genetics.* Sinauer Associates; Carroll, S. B. (2005). *Endless Forms Most Beautiful: The New Science of Evo Devo.* W. W. Norton & Company.

6. References: Bell, G. (1982). *The Masterpiece of Nature: The Evolution and Genetics of Sexuality.* University of California Press; Freeman, S., & Herron, J. C. (2007). *Evolutionary Analysis.* Pearson.

7. References: Bird, A. (2007). *Perceptions of epigenetics.* Nature, 447(7143), 396-398. Goldberg, A. D., Allis, C. D., & Bernstein, E. (2007). *Epigenetics: A Landscape Takes Shape.* Cell, 128(4), 635-638.

8. DNA damage and repair, see: Lindahl, T. (1993). *Instability and decay of the primary structure of DNA.* Nature, 362(6422), 709-715; Jackson, S. P., & Bartek, J. (2009). *The DNA-damage response in human biology and disease.* Nature, 461(7267), 1071-1078.

9. For insights into the human genome: Collins, F. S., Morgan, M., & Patrinos, A. (2003). *The Human Genome Project: Lessons from Large-Scale Biology.* Science, 300(5617), 286-290; Venter, J. C., et al. (2001). *The sequence of the human genome.* Science, 291(5507), 1304-1351.

10. Genetic engineering and its ethical implications: Doudna, J. A., & Charpentier, E. (2014). *The new frontier of genome engineering with CRISPR-Cas9.* Science, 346(6213). Caplan, A. L., Parent, B., Shen, M., & Plunkett, C. (2015). *No time to waste – the ethical challenges created by CRISPR.* Embo Reports, 16(11), 1421-1426.

From the Big Bang to Consciousness

From the Big Bang to Consciousness

From the Big Bang to Consciousness

PART TWO

THE TRANSFORMATION

From the Big Bang to Consciousness

Chapter 4

THE STRATEGY OF THE UNIVERSE FOR BALANCE

From the Big Bang to Consciousness

From the Big Bang to Consciousness

> We are survival machines –
> robot vehicles blindly programmed
> to preserve the selfish molecules known as genes.
> This is a truth which still fills me with astonishment.
>
> RICHARD DAWKINS
>
> *The Selfish Gene,* 1976

From the Big Bang to Consciousness

4.1 The mechanisms of evolution

Evolution is the fundamental process that has relentlessly shaped life on Earth, and it is regarded as the universe's strategy for creating complexity and maintaining balance. It is not a simple biological mechanism but a *pervasive force* that shapes the diversity and adaptability of life. In simpler terms, it is the process – known as natural selection – by which species change over time, making those best suited to their environment able to survive and reproduce more effectively. Yet natural selection is only the tip of the iceberg; underneath is an ocean of genetic complexity and environmental interactions. Yet beneath the apparent simplicity lies a system of mechanisms working in harmony, resulting from billions of years of natural experiments, successes, and failures in the great laboratory of the Earth. Mechanisms responsible for the myriad life forms we see all around us, from the smallest bacteria to the largest mammals, each witnessing a unique evolutionary story, a tale engraved in their DNA.

Genetic variation is the engine of the evolutionary process. Variation is life, and life, in its essence, is variety. Such variation comes from mutations, errors in DNA replication, and genetic recombination that occur during sexual reproduction. Each mutation is a new note in an ever-changing musical score, each recombination a new melody. A mutation can alter an organism's characteristics, potentially improving its ability to survive and reproduce – or negatively affecting its fitness. Most mutations are neutral; they are neither beneficial nor harmful to the organism. Nonetheless, every neutral mutation is potentially helpful for future innovation, a reservoir of possibilities from which evolution can draw, a kind of veritable reservoir of genetic variability.

Natural selection thus acts on the pool of genetic variation. It is not a conscious process of choice – instead, it is an invisible filter. Organisms that possess traits that improve survival and reproduction are more likely to pass those traits on to their offspring. In the struggle for existence, natural selection is the arbiter that rewards adaptation and resilience. Over generations, advantageous characteristics become more common within a population, leading to evolutionary change as a whole.

Change is the fabric of a story unfolding in every habitat, in every ecological niche on Earth. However, natural selection is not the only process that drives evolution. *Genetic drift*, another evolutionary mechanism, refers to random changes in the frequency of alleles (gene variations) in a population. Random changes like dice thrown by fate sometimes overturn the fortunes of an entire species. In small populations, genetic drift can lead to significant changes – independent of natural selection. Moreover, *gene* flow – the transfer of genetic variation from one population to another – can also influence evolutionary trajectories. Gene flow is the wind that carries the pollen of genetic potential across boundaries, mixing the genetic makeup of different populations. Migration, a primary source of gene flow, can introduce new genetic material into a population, like a river bringing nutrients to new soil, enriching the genetic diversity of its delta, affecting its genetic composition and potentially its evolutionary course. In an increasingly connected world, gene flow represents a promise of new combinations and evolutionary possibilities. The interplay of genetic variation, natural selection, genetic drift, and gene flow creates the diversity that results in the complexity of life we see on Earth today, demonstrating that evolution is a process of endless creativity.

Evolution unveils a profound mystery of existence, showing how – despite the universe's immensity – life finds its own particular way to emerge, adapt, and even thrive. Reflecting perhaps a broader dimension of reality that goes beyond mere survival and reproduction. Beneath the apparent chaos of evolutionary change lies a pattern, an order that challenges us to seek deeper meaning. Beyond phenotypic and genetic change lies an unceasing tide of transformation that carries the promise and potential of new manifestations of life. Life, in essence, is transformation – a transformation driven by adaptation and aspiration toward new forms of expression. *An inherent tension between the form and the formless, between being and becoming.* A beating heartbeat of evolution, the engine of a process that shapes the diversity of life in ways that continue to amaze us. In its mechanical indifference, natural selection seems to reveal a kind of inherent wisdom. This wisdom manifests itself through an invisible force that, through the veil of randomness, guides life toward ever greater complexities. Yet, in every aspect of evolution, there is a tinge of mystery, a reminder that there are aspects of life that we may never fully understand. At the same time, evolution challenges our understanding of purpose, for it does not act with a predetermined goal but proceeds through a series of unplanned and often unpredictable emergencies.

Evolution, like a river flowing through the landscape of life, continually changes the terrain, creating new paths and possibilities. And like any river, evolution has its source and its mouth, but its course is a story that continues to be written, a tale without end. But what drives this river? Is it a blind, mindless force, or is there some hidden intelligence at the heart of the evolutionary process? Could there be an intrinsic logic to evolution, an underlying matrix that defines the paths that life can take? Our understanding of evolution, then, is not only a matter of science but also touches on fundamental questions of philosophy and, ultimately, of our position in the cosmic order. In this sense, evolution is a bridge between knowing and feeling, knowledge and experience, and the concrete and the abstract.

4.2 Survival of the fittest

Survival of the fittest – the iconic phrase, often associated with Charles Darwin, actually originated with British philosopher Herbert Spencer, who sought to encapsulate the principle of natural selection. Spencer conceived adaptation as an organizing principle, a force that shapes society as well as biology. Although the phrase has often been misunderstood to mean *survival of the most aggressive*, in the context of evolution, the term "adapted" refers to an organism's *fitness* for its environment in terms of reproductive success. Suitability, or *fitness*, mainly when studied in higher species, is not measured by physical strength or aggressiveness but by an organism's overall ability to fit harmoniously into its ecosystem and includes an extensive range of qualities, physical, social, and even aesthetic. From a thermodynamic perspective, the concept of fitness can be interpreted as the ability of an organism to harness and utilize energy in its environment effectively.

Organisms with high *thermodynamic fitness* are masters in the art of survival, turning available resources into reproductive success.

Every process of life, from the most straightforward cellular function to complex behaviors, requires energy. Life is a constant energy conversion cycle – from a star's core to an ocean's depths. The laws of thermodynamics govern energy acquisition, allocation, and dissipation. Laws are not just physical rules but principles that creatively define the possibility of life. The organisms that manage thermodynamic processes most effectively in the competition for survival are usually the most successful. Thus, evolutionary success is a matter of balance, not mere dominance.

For example, the cheetah's ability to run at extraordinary speeds to catch prey manifests its thermodynamic fitness. The cheetah embodies efficiency, an evolutionary design model that optimizes energy use for survival. The energy it extracts from food is efficiently converted into kinetic energy, merging physics with biology in the act of hunting, enabling it to secure the next meal and thus survival and reproductive success. Similarly, a tree's ability to harness sunlight for photosynthesis is another example of *thermodynamic fitness*. Trees are pillars of ecosystems, solar transformers that power life on Earth. By converting solar energy into chemical energy, trees fuel their own growth and reproduction and create energy-rich resources (fruits, nuts, leaves) that support other organisms, serving as central nodes in the web of life, thus shaping the energy flow of entire ecosystems.

In particular, the measure of an organism's fitness is not only its individual survival but its success in *transmitting information* to the next generation. I deliberately use the formula "transmit information" and not "transmit genes" because sexual reproduction is a subset of the broader concept of transmission of information along the arrow of time. The work of Aristotle, Michelangelo, and Shakespeare influences the future of humanity far more than their direct genetic transmission. And what about Johannes Gutenberg, the inventor of the printing press, who brought about a revolution in the transmission of information far beyond its biological hereditary transmission? That is why nature has appropriated such a mechanism, often full of what appear to be sacrifices on the surface but are actually investments in the future.

Let us return to the animal sphere. Consider the example of the male peacock's extravagant tail. The peacock's tail is a masterpiece of visual communication, a calling card for sexual selection. Although it requires considerable energy to produce, maintain, and potentially attract predators, it is living proof of the peacock's fitness. It signals health and genetic quality to potential mates in the *form of image* or even *aesthetic sense*, improving the peacock's chances of securing a mate and passing on its genes. The tail is thus more than just an appendage; it is a statement of evolutionary success. But beware: the tail is not an efficient structure in a mechanical sense; it is so in its ability to arouse attention, opening up new dimensions, including those of a *symbolic nature*. Beauty in the animal kingdom is not merely ornamental; it is a language, a way of communicating fitness in an evolutionary context.

The principle of *survival of the fittest* is thus closely linked to energy flow but produces results that transcend mere physical functionality. Fitness is an interweaving of functional, aesthetic, creative, and even ethological (*animal ethos* manifested in the totality of its behaviors) traits, which together define an organism's ability to thrive in its environment. The phrase "survival of the fittest" evokes deep reflection on the very meaning of adaptation and survival in a universe in constant flux. By illuminating the mechanisms of life, evolutionary theory confronts us with an essential question: *what does it mean to be genuinely adapted in a world where change is the only constant?* The answer to that question forces us to peer beyond the surface of things, beyond the pure materiality of being and the principle of pure dominance.

In an uncritical analysis, the concept of survival of the fittest could be interpreted as a legitimization of cutthroat competition, an echo of the law of the jungle where only the fittest or the most skilled deserve to survive and thrive. Such an interpretation would reduce evolution to mere contention for scarce resources, ignoring the richness of cooperation, symbiosis, and interdependence that permeate life. Indeed, the very nature of adaptation evokes an idea of *creative response to challenges*, an openness to the new, to the unfamiliar, challenging the very notion of adaptation as a mere functional arrangement. The exuberance of the peacock, the harmony of a flock of birds in flight, or the complexity of a colony of corals all speak to a dimension of adaptation that goes beyond mere survival-an *aesthetic celebration* of existence, an expression of nature's inherent creativity. Adaptation becomes an act of *creative participation* in the larger drama of existence, an invitation to explore new possibilities of *being in the world*.

The notion of *fit* in the evolutionary context evokes reflection on value and meaning. *Who or what decides what is suitable?* More importantly, *fit for what?* We are faced with a narrative that could prove reductive if interpreted rigidly. Perhaps, beyond the physics of evolution, there is an *evolutionary metaphysics* of being waiting to be explored, a realm where the categories of fit and unfit dissolve into a deeper understanding of life as an inexhaustible mystery that continues to unravel across time and space. As conceived and understood, the discourse of survival of the fittest opens the door to a deeper dialogue between science, philosophy, and ethics, inviting a broader reflection on humanity and its place in the larger context of cosmic evolution.

4.3 Evolution as energy optimization

A unifying theme in life history is optimization – the fine-tuning of form and function to maximize efficiency – which we have since seen to be efficient in the diffusion of information along the arrow of time. However, *the question of efficiency in the service of what end remains elusive and entirely open*. The relentless pursuit of information efficiency is the underlying engine of every evolutionary adaptation, from the leaf that twists to catch sunlight to the predator that sharpens its senses to hunt better to the writer who immortalizes eternal verses. In its material essence, life can be seen as a dynamic process of transformation and optimization of energy – a chaotic transformation governed by principles that seek harmony between energy needs and environmental possibilities.

Every organism, cell, and biomolecule is a conduit for energy flow, directing it along pathways that enable growth, reproduction, and survival. It is an actual economy of nature, where energy is the currency and evolution is the market. Evolution is the process that shapes the pathways, driven by the relentless pressures of natural selection. It acts as a strict accountant, favoring energy balances that promise greater returns.

In simpler terms, energy optimization is about maximizing the conversion of available energy into useful work. Practical work is not just mere survival but also the ability of an organism to leave behind an offspring, a copy of itself, in the infinite flow of time. An organism that is more efficient in extracting energy from the environment and converting it into growth and reproduction has a selective advantage over its less efficient counterparts. Selection in favor of efficiency seems to be a principle found throughout the natural kingdom.

Consider the evolution of photosynthesis, one of the most crucial processes on Earth. The ability of plants to capture sunlight and convert it into chemical energy (glucose) is an extraordinary feat of energy optimization. Plants are pioneers of renewable energy, nature's first and most efficient solar panels. It supports plant growth and reproduction and powers most life on Earth. Such is the gift of photosynthesis: life feeding on light, turning the Sun into nutrients! However, photosynthesis, as we know it today, did not appear in one sudden step; it resulted from a slow and gradual evolution that began with early organisms using simple and inefficient processes to capture energy. Only over time, with natural selection favoring organisms, we were able to develop more efficient ways of capturing and storing sunlight that we arrived at the highly efficient process of photosynthesis that we observe in modern plants. It is an epic of incremental improvement, a tale of refinement spanning a distance that can be measured in hundreds of millions of years. Another striking example of energy optimization in evolution is the development of *endothermy* (warm blood) in mammals and birds. The evolution of the internal thermostat is an example of how evolution refines the tools of life to thrive in a broader range of environments. Endothermic animals can optimize metabolic reactions by maintaining a constant body temperature, allowing high activity levels even in cold environments. Internal regulation is an act of independence from environmental fluctuations. Although endothermy requires a significant investment of energy, the benefits in terms of survival and reproductive success can outweigh the costs, resulting in persistence over evolutionary time. It is a gamble on life, where the highest stakes can lead to the most profitable rewards.

There is a silent invocation, a call for reflection on the balance between efficiency and vitality, between function and form, between being and becoming. The narrative of energy optimization can also be seen as a metaphor for the human search for meaning and perfection. As organisms evolve and improve through time, so does our understanding of the world and our place in it progress. A relentless search for essence, a drive toward enlightenment that, in many ways, reflects the optimization of energy in nature. *Evolution is not only a physical process but also a metaphor for our spiritual and intellectual journey.*

On the other hand, the notion of energy optimization raises a critical question: in a world obsessed with efficiency, what is lost along the way? Is efficiency a value in itself, or is it a means to a higher end? In an era where economic and technological efficiency has changed from a means to an end, the questions are particularly poignant. *Efficiency without the guidance of higher values can lead to alienation, destruction, and exploitation.* Beauty, joy, love, and art – aspects so linked to the search for meaning in human life – escape the measure of efficiency yet remain central to our existence.

Adaptation and optimization are not simply mechanical reactions to external pressures but testimonies to life's creativity and ingenuity. Nature's extraordinary ability to find pathways to efficiency underscores a deeper order, an unfathomable wisdom that pervades existence and calls for a multidimensional inquiry that goes beyond the material and points to a more holistic and integrated understanding of evolution.

4.4 The role of mutation

Mutations are the raw material of evolution; they introduce variations into the *gene pool* on which natural selection acts. But how do mutations occur, and what role do they play in the grand design of life? At the molecular level, mutations are alterations in the DNA sequence, the *blueprint* of life. Alterations that can be likened to unexpected notes in a symphony are often dissonant but sometimes *give rise to a new melody*. Changes can occur through various mechanisms, such as errors during DNA replication, exposure to radiation or chemicals, or random events. Mutations that, though minor, can have a huge impact, acting like the dice of fate in reshuffling the code of life.

Most mutations are neutral and do not significantly affect an organism's *phenotype* (observable characteristics). Some can be deleterious, causing harmful effects and turning the mutation into a toll that life pays for its diversity, while others can be beneficial, providing the organism with a selective advantage. *Beneficial mutations* are like winning tickets in an evolutionary lottery: rare but *transformational*.

From a thermodynamic perspective, mutations introduce an element of randomness or *noise* into the system. Noise is nothing more than nature's creativity at work, the uncertainty from which novelty arises. Noise can be seen as an increase in entropy at the molecular level. In its ingenuity, life uses entropy not as an obstacle but as an opportunity: it is precisely the increase in entropy that fuels the diversification and complexity we observe in the macroscopic world of living organisms.

Evolution, then, can be considered an orchestra that intones the melody of life, with mutation as one of its boldest and most innovative composers. While a single mutation is unlikely to have a significant effect, the cumulative impact of multiple mutations over generations can lead to substantial changes in an organism's characteristics. Cumulative mutations are the brushes with which nature paints the evolution of *biodiversity*.

Suppose changes improve an organism's ability to survive, reproduce, and efficiently use energy in its environment. In that case, they can become more prevalent in the population through natural selection, driving the process of evolution. This demonstrates how even the most random events can be forged into tools for adaptation and survival.

For example, consider the evolution of antibiotic resistance in bacteria. A mutation that enables a bacterium to survive exposure to an antibiotic gives it a significant advantage. Its mutation is an adaptation, a survival tactic in an evolutionary war that takes place on a microscopic scale. Although mutation may initially seem like a random, entropic event, its result leads to a population of bacteria better adapted to their environment, illustrating how evolution continually refines life through trial and error.

Moreover, mutations that may be harmful or neutral in one context may turn out to be beneficial in another, depending on environmental conditions, in a phenomenon known as *environment-dependent fitness*. This further underscores the interplay between mutation, natural selection, and energy flow dynamics and reminds us that *fitness is a relative, not an absolute, measure.*

The *dialogue between the known and the unknown* is an eternal narrative of human existence. The tendency to explore, discover, and overcome new horizons is embedded in the very essence of being human. Mutations, in the context of evolution, are a material manifestation of this narrative, representing a step into the unknown and an opening to new possibilities. They are the seeds of change, the sparks of biological creativity that can potentially drive life's evolution into unexplored domains. The *courage to face the unknown*, embodied in the mutational process, is a quality that resonates deeply with the essence of human exploration. When a mutation emerges, it brings the potential for new forms of life expression, new solutions to old problems, and new responses to challenges posed by the environment. A process that is a reflection of the universe's relentless drive toward complexity and diversity.

Likewise, every human innovation, every new idea or discovery, is a step into the unknown, a plunge into the realm of limitless possibilities. Reflection on mutations also leads to a more profound reflection on the intrinsic value of individuality and diversity (and its defense). *Each mutation is a unique experiment*, and individual deviations from the normal represent the source of life's diversity. In a world that often prizes conformity and fears the unknown, the story of mutation reminds us of the value and importance of individuality. Just as mutations enrich the gene pool, different perspectives and experiences enrich our understanding of the human experience.

Moreover, the role of mutations in evolution challenges the idea of static perfection. Perfection, in the context of evolution, is relative and dynamic. What is perfect in one context or moment may not be ideal in another. Thus, evolution, driven by mutation and natural selection, proposes a view of perfection as a continuous process of adaptation and growth rather than a fixed state.

On the broader stage of human life and evolution, mutations are evidence of impermanence and the constant flow of change. As human beings, we are called to navigate the changing waters of the unknown, embracing the opportunities for learning, growth, and transformation that come with uncharted territory. And while initially, mutations may seem random or even harmful, only to turn out to be beneficial, so too can the challenges and obstacles we encounter along the way serve as catalysts for personal and collective growth and transformation. In this way, the role of mutations in evolution is not just a biological discussion but an invitation to reflect on the nature of change, uncertainty, and opportunity in the context of human experience.

4.5 The rise of complexity

The trajectory of life's evolution has been marked by an undeniably apparent trend: increasing complexity. From the single-celled organisms that dominated the Earth for billions of years, life has branched out to today's astonishing diversity, encompassing intricate organisms, sophisticated behaviors, and dynamic ecosystems. This rich and varied mosaic of life testifies to nature's ability to experiment and refine, to expand the frontiers of existence into ever more elaborate forms.

But how do we reconcile the emergence of complexity with the Second Law of Thermodynamics, which mandates a universal drift toward disorder? From a thermodynamic perspective, complexity can be seen as a sophisticated energy flow and dissipation mechanism. The dialectic of building and destroying order takes place in the context of a universe that, as a whole, drifts toward a state of increasing disorder. Although it may seem counterintuitive, the growth of complexity does not contradict the Second Law; rather, it represents a strategy that life has developed to persist and thrive in a universe dominated by entropy.

Organisms create order within their structures by exporting entropy to their surroundings, often in the form of heat in a balance between creation and dissipation that is not an accident, of course, as much as a fundamental process that defines life itself. Complexity can be seen as process optimization, a means of maximizing energy use and dissipation while minimizing waste. Life's strategies to achieve this goal are as varied as they are surprising. Consider multicellular organisms, for example. They represent a qualitative leap in energy management, an innovation that has enabled a division of labor and specialization that is the epitome of biological complexity. A multicellular entity can harness different energy sources, survive in various environments, and generally manage resources more efficiently than a collection of independent cells. By developing specialized cells and organs, multicellular organisms can optimize energy allocation and utilization, leading to greater thermodynamic efficiency.

Similarly, complex behaviors and social structures – from flocks of birds to human societies – can be viewed as strategies for improving energy use. They enable individuals to protect resources, share information, coordinate activities, and ultimately enhance their survival and reproduction, which is the goal of evolutionary success. Yet, *complexity is not always the preferred path*. Evolution is opportunistic and favors solutions that offer immediate benefits for survival and reproduction. Complexity comes at a cost: it requires more energy to maintain and can increase vulnerability to specific threats. Therefore, in some environments and for some organisms, simpler structures and strategies may prevail.

Complexity, as such, is an odyssey reflecting life's relentless drive toward expansion, discovery, and realization of unexplored potential. It reflects life's innate curiosity and insatiable hunger for exploration and adaptation. Yet, new challenges, vulnerabilities, and questions emerge with each new degree of complexity. Complexity is an invitation to a deeper understanding of the nature of existence, meaning, and value.

The rise of complexity can be seen as an expression of life's inherent desire to explore, create, and become. An echo of the insatiable curiosity, indomitable ingenuity, and inexhaustible creativity that are at the heart of the human experience. The concept of complexity also evokes reflection on the nature of progress and its cost. Progress, represented by increased complexity, carries promise and danger with it. And, just as biological complexity requires greater expenditure of energy and resources, so too does human progress often require progressively more significant investments of time, energy, and resources, with no promise of return on investment, thus explaining why excess complexity from living organisms to society and civilizations, can often culminate in *systemic implosion*.

4.6 The Cambrian explosion

About 540 million years ago, Earth experienced an extraordinary event known as the *Cambrian* explosion, which marked a relatively rapid emergence of a wide range of complex multicellular organisms, a period that represents a significant milestone in the history of life, signaling the arrival of nearly all the major animal groups we see today. The Cambrian explosion can be interpreted as a period of ecological exploration, where life experimented and diversified its survival strategies through an exceptional acceleration in the rate of biological innovation.

How can we interpret the incredible explosion of biodiversity and complexity? The Cambrian explosion represents an example of how energy and information combine to overcome the universe's entropic tendency toward disorder.

The Cambrian explosion coincided with a period of significant geological and climatic change. The Earth was warming after a global glaciation event, increasing atmospheric oxygen levels and creating favorable conditions for the evolution of larger, more complex life forms in a phase that represented a convergence of factors that paved the way for evolutionary innovation. Increased oxygen levels enabled the evolution of aerobic respiration, which allows organisms to extract significantly more energy from food than anaerobic processes, resulting in energy optimization crucial to the evolution of biological complexity.

The Cambrian biosphere acted as a laboratory in which natural selection tested countless solutions to meet the challenge of harnessing newly available energy resources in a habitat where the increased energy yield could support the growth and maintenance of larger and more complex organisms, leading to the development and evolution of a myriad of multicellular life forms. Greater resource availability also likely played a role in driving biodiversity during this period. As new ecological niches opened up, organisms evolved to exploit them, each finding a unique way to harness and dissipate energy in a diversification process that further increased the overall energy flow through the biosphere, contributing to life's ongoing challenge to entropy. With all this, the Cambrian explosion was not a one-way street to increased complexity. The variety of life forms that emerged during that period created fertile ground for the phenomenon known as *co-evolution*, in which the appearance of a new species often forced others to adapt or innovate in order to survive. This meant that as new organisms evolved, they went on to create new challenges and pressures on each other. Predators evolved with prey, parasites, hosts, competitors, and mates, each pushing the other to adapt, innovate, and optimize their energetic strategies in a dynamic and ever-changing life landscape. Dynamism and interdependence made life a creative process, a relentless leap forward in complexity and diversity.

It was a time when life seized a series of favorable conditions to expand its reach massively, explore new pathways of energy flow, and create a surprising array of forms and functions. The Cambrian explosion demonstrates that *periods of crisis can become moments of evolutionary opportunity*, propelling life to reach new levels of complexity and specialization. The echoes of the ancient explosion continue to resonate in the diversity and complexity of life we see today, underscoring the resilience and ingenuity of nature, which is constantly finding new ways to adapt, thrive, and overcome environmental challenges.

Thus, the Cambrian explosion represents an emblem of possibility, a shining moment in the vast period of evolution. The veil of existence seemed to have given way, revealing a domain of infinite forms and manners of being. In the immensity of its creativity, life seemed to have broken the chains of simplicity, unleashing a kaleidoscope of biological fantasies.

4.7 Evolutionary arms races and energy dissipation

Life on Earth is characterized by continuous interaction and co-evolution among species, often leading to what biologists describe as an *evolutionary arms race*. Predators improve their hunting strategies and weapons, prey responds by enhancing their defenses, parasites evolve to exploit their hosts better, and hosts adapt to resist their parasites. Arms races that appear as endless cycles of adaptation and counter-adaptation, where fitness is determined not only by the ability to survive but also to reproduce in a given environment.

Where does thermodynamics fit into this dynamic picture? Evolutionary arms races, as manifestations of co-evolution, are dynamic scenarios where the complexity of life is observed to unfold in cycles of adaptation and counter-adaptation and increase its biodiversity and complexity of survival strategies. Each advanced hunting strategy or defense mechanism is a thermodynamic solution for the organism, which seeks to maximize efficiency in converting available energy into useful biological work, whether to capture prey or escape a predator. Entropy is, in fact, not only a measure of disorder but also an index of energy flow and *information transfer* within and between organisms. Interestingly, genetic information encoded in DNA can be viewed in the light of information theory, which is intertwined with thermodynamics in the context of evolutionary biology.

The arms races represent an expression of the ebb and flow of energy and act as an engine for innovation and complexity, continuously operating within the boundaries imposed by the laws of thermodynamics. Acting as a regulator, the process of natural selection ensures the persistence of only those adaptations that increase an organism's net energy efficiency, thereby contributing to its evolutionary fitness.

The availability of energy resources and efficiency in their use manifest themselves in various ways. For example, organisms that develop less energy-intensive strategies to maintain vital functions enjoy an advantage in environments where energy is scarce. In contrast, energy-intensive strategies may not be sustainable in the long term, especially if they do not offer significant benefits in terms of survival or reproductive success, a reminder of the *principle of energy economics* that highlights the tendency of evolution to favor solutions that optimize resource use.

The cycles of adaptation and counter-adaptation illustrate life's resilience and ingenuity in confronting and adapting to an ever-changing environment, always respecting the fundamental laws that govern our universe. Such a process can be seen as a metaphor for psychology and, more generally, the human condition, where the individual is constantly engaged in a journey of personal adaptation and growth, influenced by both external forces and one's own internal decisions.

4.8 The emergence of intelligence

Intelligence is a fascinating aspect of the biological world, manifesting particularly strikingly in humans and varying wildly among different animal species. The evolutionary convergence of intelligence in other species suggests that it is an extremely advantageous attribute, such that different environmental pressures may favor its evolution independently. This convergence phenomenon is eloquent evidence of how intelligence offers adaptive solutions to similar ecological challenges, reinforcing the idea that different systems can reach similar conclusions through distinct evolutionary paths.

Organisms capable of learning from their experiences, anticipating outcomes, and making informed decisions gain a significant advantage for their survival. This ability to learn and anticipate is crucial not only at the individual level but also extends to the profound implications for the collective progress of species. The question remains: how can we interpret the emergence of intelligence through the thermodynamic perspective? At a more fundamental level, intelligence can be understood as a sophisticated strategy for managing energy and minimizing entropy. From a thermodynamic perspective, it could be seen as a mechanism to increase efficiency in energy use, reducing entropy at the local level while contributing to the overall increase in entropy of the universe. Nevertheless, the evolutionary benefits of intelligence come at a significant energy cost.

The brain, the hub of intelligence, is a metabolically onerous organ. Although it accounts for only about 2 percent of body weight, it consumes between 20 and 25 percent of the body's energy. The relationship between the brain's weight and its energy consumption clearly illustrates how evolutionary pressures can influence organisms' metabolic strategies. In particular, the human neocortex has undergone a disproportionate expansion compared to other species, suggesting the presence of a biological basis for advanced human intellectual capabilities. The expansion of the neocortex is one of the most evident signs of how evolution tends to favor improvements in cognitive abilities, even as energy expenditure increases. Maintaining such a high energy expenditure requires a constant supply of nutrients, which implies the development of effective strategies for locating and acquiring food.

Agriculture, animal husbandry, cooking, and food preservation are all examples of how culture, unique to humankind, enables organisms to optimize their behavior, thereby improving energy acquisition and survival over time. Such behavioral optimization manifests itself in cooperation and competition, essential elements of social interactions, both of which are influenced and shaped by intelligence. By learning how to use tools, early humans were able to access new food sources, such as marrow within bones, acquiring the energy needed to sustain large, energy-hungry brains. The invention and use of tools highlight the presence of intelligence and advanced planning, paving the way for new forms of culture and civilization. The complexity of human social structures may have stimulated the development of advanced cognitive capabilities functional for managing intricate social interactions. Indeed, human societies are distinguished not only by their complex structure but also by institutions that reflect and further promote the development of individual and collective intelligence.

By developing such advanced cognitive capabilities, life has discovered a method to channel the energy flow, facilitating the creation of increasingly elaborate structures and behaviors. The *process of cognitive sophistication* shows that evolution is not only about physical changes but also includes intellectual and cultural development. Because of this, life can anticipate and respond to the future, strategize and plan, shaping its environment in previously unimaginable ways. The ability to anticipate and plan are distinctive features of intelligence that distinguish cognitively advanced organisms from those that rely solely on pre-programmed instincts and reactions.

Neuroplasticity – the *brain's ability to reorganize itself by forming new neural connections* – is a crucial element of intelligence. It facilitates adaptation to new experiences, learning new information, and healing from trauma. This phenomenon highlights the brain's exceptional ability to adapt and restructure itself in response to learning and environmental stimuli.

Thinking, reasoning, learning from the past, and planning for the future constitute forms of mental order beyond organisms' simple physical organization. Reflection, abstract reasoning, and planning indicate a higher level of awareness and control over one's surroundings. Intelligence acts as a bulwark against the uncertainty and unpredictability of existence, equipping organisms with the necessary tools to navigate a sea of variability with a compass of understanding. The ability to navigate consciously becomes essential for resilience and behavioral flexibility in an ever-changing world. Moreover, through its capacity for reflection and conception, intelligence introduces a new dimension to life: *mindfulness*. It extends from simple self-awareness to the ability to ponder existential and metaphysical questions, a defining characteristic of human beings.

Intelligent organisms do not merely react to the world around them; they reflect on it, imagine it, and try to change it following a vision or desire. Creativity and innovation emerge from reflective capacity and are fundamental to technological and cultural progress. The ability to imagine and create marks a kind of *awakening*, a transcendence from being merely subject to the forces of nature to becoming agents capable of influencing, at least in part, one's own destiny. Consequently, intelligence can be seen as an evolutionary mechanism that promotes freedom, self-determination, and the creation of meaning.

However, with high capacity also comes great responsibility. The broad power that intelligence confers also imposes a moral duty to use it in ways that will benefit not only the individual but also society and the ecosystem as a whole. With its power to modify the environment and influence other life forms, it carries a duty to act with wisdom and consideration. It is critical that we, as an intelligent species, recognize and respect the interdependence between all living things and the shared environment. Human history offers numerous examples of how, if misused or ignored, intelligence can cause destruction and suffering on a large scale.

The concept of intelligence is closely related to the philosophical debate between *free will* and *determinism*. This debate sheds light on whether our actions derive from an inevitable causal chain or whether we possess the ability to choose our own path independently. The ability to plan, strategize, and make decisions may indicate a degree of behavioral autonomy that contradicts deterministic views of the universe, making free will a pillar of our conception of personal and moral responsibility.

Moreover, a deeper reflection on the essence of intelligence emerges. Despite its apparent superiority, it remains rooted in the thermodynamic ecology of our world, reminding us that despite our intellectual achievements, we are still subject to the laws of physics and ecology. Intelligence is ultimately a means of navigating the ebb and flow of energy and entropy. Our need to understand and manipulate the environment stems from the same natural laws governing all life forms' survival. Despite its greatness, intelligence does not elevate us above our fundamental nature as living beings in a physical universe. It is a reminder of how our every thought and invention is deeply intertwined with the very fabric of reality. Therefore, intelligence is not only a means to overcome the pragmatic problems of survival but also a gateway to a deeper understanding of our existence, a window through which we can interrogate the very meaning of being. In this way, intelligence is revealed as an *evolutionary gift* that enables active participation in exploring and understanding the vastness of life and existence.

4.9 Human evolution

From our humble beginnings as tiny tree-dwelling primates to our rise as the planet's dominant species, human history vividly illustrates life's challenge to entropy. About seven million years ago, our ancestors began to diverge from those of our closest living relatives, the chimpanzees and bonobos. Paleoanthropology informs us that this divergent path is marked by important fossil finds, such as the famous *Lucy*, which provide valuable insights into the human evolutionary process.

The transition from a tree lifestyle to a terrestrial one opened up new possibilities and challenges, reshaping our ancestors' relationships with the environment and each other. Bipedalism may have facilitated the display of predators or prey and the release of hands for carrying food and children. Over millions of years, our lineage has undergone a series of environmental changes, from forests to grasslands, from tropical to temperate climates, and from isolated groups to global interconnectedness. These changes are evidenced by variations in fossils and material cultures found by archaeologists. Each transition required our ancestors to adapt their energy strategies, affecting our physical characteristics, behaviors, and social structures. Diet, for example, evolved to include more energetic foods such as cooked meat and tubers.

Perhaps the most significant event in human evolution was the development of *bipedalism*, the ability to walk on two legs. Biological anthropology suggests how bipedalism improved energy efficiency by reducing sun exposure and caloric expenditure – a change with profound thermodynamic implications. Bipedalism is a more energy-efficient mode of locomotion over long distances than the quadrupedal movement used by our primate relatives. Such efficiency has been essential to our survival during periods of climate change and migration. Greater efficiency probably allowed our ancestors to expand their areas of residence, explore new environments, and exploit a wider variety of food sources. The *geographic dispersal* of humans can be traced through genetic and archaeological *markers* that map our journey out of Africa. The development of bipedalism has not only altered how individuals move through the physical world but also freed the hands, enabling a range of interactions and manipulations that have further catalyzed technological and cultural innovation. The liberation of the hands led to the development of fine manual skills essential for creating and using tools. The advent of tool use and the control of fire are other milestones in our evolutionary journey, representing new ways of harnessing and manipulating energy. Stone tools, for example, enabled the slaughter of large animals, increasing protein and fat intake in the human diet.

Fire provided a means of modifying the environment, offering protection, warmth, and the ability to cook food. Fire manipulation has also had a social impact, creating a *convergence point for meeting and sharing knowledge*. Cooking, in particular, has profound thermodynamic benefits as it helps break down food, reducing the energy required for digestion and increasing the net gain of energy storage. Cooked food improved nutrient absorption and supported the growth of our large, energy-hungry brains. The adoption of tools and the control of fire represent eloquent examples of how human ingenuity has expanded our ability to interact with and manipulate the environment despite the restrictions imposed by entropy. Such developments have driven innovation in other fields, such as clothing and shelter, increasing our ability to survive in previously inhospitable environments. They are an expression of life's urgency to adapt, innovate, and transcend immediate limitations.

Our large brains, a defining characteristic of our species, represent a significant energy investment. As we discussed earlier, energy-hungry organs confer substantial advantages in survival and reproduction, enabling advanced problem-solving, communication, social coordination, and cultural transmission. Advanced cognitive skills have been crucial to developing everything from language to art, science to philosophy. Cognitive abilities enabled early humans to form complex societies, develop technology, and radically alter the earth's biosphere.

Our influence on Earth is so profound that some scientists believe we have begun a new geological era: *the Anthropocene*. As philosopher Daniel Dennett has clearly stated, the evolution of consciousness and intelligence represents a leap into evolutionary freedom, where culture and knowledge become evolutionary forces parallel to natural selection. In many ways, human evolution is the story of how a primate lineage managed to carve out an ever-widening thermodynamic niche for itself, relentlessly expanding its ability to harness and dissipate energy. Our species has shown a unique ability to alter its environment in ways that amplify our evolutionary success, from the development of agriculture to the industrial revolution.

The history of human evolution is thus a testimony to how life, despite thermodynamic restrictions, has found ever more creative and ingenious ways to persevere and thrive. Each innovation, each adaptation, represents a kind of challenge between the creative essence of life and the physical laws that govern the universe. As we reflect on our evolutionary past and look to the future, we are invited to contemplate our earthly roots and the potential of *creativity in the face of the unknown*. Such philosophical reflection prompts us to consider the future in terms of what we can do and what we should do, *balancing innovation with wisdom and introspection*.

4.10 The future of evolution

The current life trajectory on Earth is being shaped by forces that our ancestors could hardly have imagined. Rapidly advancing technology, growing populations, and the profound changes we are making to the planet's ecosystems create a complex and dynamic landscape that future evolution will traverse. For example, current climate and *biodiversity* challenges are already influencing evolutionary directions in ways that traditional ecology might not have predicted.

One thing is sure: as long as life exists, it will strive to persist and multiply in the face of entropy, creatively finding new ways to harness and dissipate energy. This concept of resilience is highlighted in the *panspermia* hypothesis, suggesting that *life, once born, can spread through the universe* with specific paths that evolution will take and the forms that life will take – largely unpredictable.

As philosopher and biologist Francisco Varela reminds us, life exhibits a tendency toward *autopoiesis*, a capacity for self-organization and renewal that can head toward simple or complex forms. Despite anthropogenic pressures, the resilience of ecosystems such as rainforests and coral reefs demonstrates autopoiesis in action.

In contemplating the end game of entropy, we might ask: will the future of evolution lead to ever-increasing complexity? Not necessarily. As we have seen, complexity is *one* strategy that life uses to manage energy and entropy, but *it is not the only* strategy, as much as it is not always the best one. The past and potentially future phenomenon of *mass extinction* and subsequent *evolutionary radiation* show that change is the only constant. Simple organisms like bacteria and viruses continue to thrive alongside much more complex ones.

Depending on specific environmental pressures and opportunities, the future could see an increase in complexity in some areas of life and a decrease in others. For example, *synthetic biology* could lead to the creation of life forms optimized for specific functions, reducing complexity in favor of efficiency.

Humans have become a significant force in shaping the course of evolution, both for us and for other species. As we increasingly assume the role of *planetary force*, we must ask: How will our actions affect the future of evolution? What responsibilities do we have to other species and future generations? Our activities may have triggered the sixth mass extinction, raising new questions about *biodiversity* conservation and *engineering*. Technology allows us to manipulate our own biology and that of other organisms in ways that were previously the exclusive domain of natural selection, and these pressing questions must be placed at the center of the debate about biotechnologies, such as *gene drive*, that could eliminate or change entire species.

At the opposite end of the doomsday and apocalyptic scenario, the convergence of technologies such as *gene editing, artificial intelligence, and nanotechnology could open up entirely new evolutionary scenarios.* Growing complexity, a dominant narrative of our evolutionary journey, could continue to unfold, propelled by the forces of technology and innovation. As philosopher and theologian Pierre Teilhard de Chardin contemplated, evolution could be seen as an upward spiral, where complexity and consciousness co-evolve in a creative dialogue with the ever-changing environment. The concept of co-evolution suggests a future where artificial intelligence and biology merge, opening new avenues for evolution.

Our ability to reflect and manipulate nature gives us a significant, if not unique, role in shaping the course of evolution. From responsible *guardians of the biosphere* to potential *architects of new life forms*, our relationship to evolution is radical and profound. The debate over geoengineering and responses to climate change exemplifies the choices we must make as *geophilosophers* – terms coined by Jan Zalasiewicz to describe our new role on the planet. Philosopher Hans Jonas, with his *imperative of responsibility*, challenges us to consider the ethical implications of our actions in a world where technology could rewrite the rules of evolution. An imperative that urges us to consider not only technical capabilities but also the wisdom that is needed in using them so that our evolutionary impact is not only powerful but also wise and beneficial

Notes and insights to Chapter 4

1. Darwin, Charles. *The Origin of Species* (1859): In his groundbreaking book, Darwin introduced the concept of natural selection and explained how species adapt and change over time.

2. Dobzhansky, Theodosius. *Genetics and the Origin of Species* (1937): Dobzhansky integrates genetics with evolution, emphasizing how mutations and genetic recombination contribute to biological diversity.

3. Mayr, Ernst. *Populations, Species, and Evolution* (1970): An influential work discussing gene flow and genetic drift in the context of populations and species.

4. Gould, Stephen Jay. *The Structure of Evolutionary Theory* (2002): Gould explores the mechanisms of evolution, emphasizing the role of natural selection combined with genetic drift.

5. Dunbar, R.I.M. *The social brain hypothesis* (1998): Dunbar proposes that increased brain size in humans is related to the need to manage complex social relationships, emphasizing the link between social intelligence and survival.

6. Mithen, Steven. *The Prehistory of the Mind: A Search for the Origins of Art, Religion and Science* (1996): Mithen explores how human intelligence developed in response to environmental and social needs, offering insight into how cognitive abilities influenced our cultural evolution.

7. Hawkins, Jeff. *On Intelligence* (2004): Hawkins discusses the neural mechanisms that support intelligence, proposing a theory of how the brain perceives, imagines, and acts in the world based on an internal predictive model.

8. Leakey, Richard, and Lewin, Roger. *Origins Reconsidered: In Search of What Makes Us Human* (1992) explores how fossil finds, including the famous Lucy, offer detailed insights into human evolution. It highlights crucial transitions such as bipedalism and their thermodynamic significance.

9. Johanson, Donald and Edgar, Blake. *From Lucy to Language* (1996): Johanson provides a comprehensive analysis of fossils and ancient human adaptations, discussing how bipedalism and other physical changes allowed humans to exploit new ecological niches.

10. Wrangham, Richard. Catching Fire: *How Cooking Made Us Human* (2009): Wrangham explores how the discovery of fire and the development of food cooking had a fundamental impact on human diet and brain evolution, with significant thermodynamic implications.

11. Dennett, Daniel. *Darwin's Dangerous Idea: Evolution and the Meanings of Life* (1995): Dennett discusses how the evolution of consciousness and intelligence has opened new horizons for human evolution, transforming the principles of natural selection into cultural forces.

12. Pierre Teilhard de Chardin (1881-1955): A theologian and philosopher, Teilhard proposed the notion of evolution as an *upward spiral* in which complexity and consciousness develop jointly. This concept is explicitly formulated in his works, such as *The Human Phenomenon* (1955), where he explores the link between physical and spiritual evolution.

13. Jan Zalasiewicz: A geologist and author, he introduced the term "geophilosopher" to describe humanity's active role in geologically shaping the planet. His research is documented in several scholarly publications concerning the Anthropocene, as detailed in his book *The Earth After Us* (2008), which examines the geological traces humanity may be leaving behind.

14. Hans Jonas (1903-1993): A philosopher known for his work on the ethics of technological responsibility, Jonas explores the moral responsibilities of technological innovation in his major work, *The Responsibility Principle* (1979). In it, he discusses the ethical implications of human actions for the natural environment and future life.

From the Big Bang to Consciousness

From the Big Bang to Consciousness

Chapter 5

Memory:
A thermodynamic picture of the experience

From the Big Bang to Consciousness

Great is this power of memory,
Too big, my God,
A vast, infinite sanctuary.
[...]
And yet it is a faculty of my spirit,
connected to my nature.

ST. AUGUSTINE

Confessions
(Book X, Chapter 8), 397-400 AD.

From the Big Bang to Consciousness

5.1 The nature of memory

Memory, the ability to store, retain, and retrieve information, is one of the brain's most fascinating abilities. It is central to our identity, learning, and decision-making, providing a *continuum* between past experiences and future actions. Despite its everyday familiarity, the processes underlying memory formation, storage, and retrieval are complex and, therefore, a central object of research for several decades. In its multifaceted essence, it unfolds not only as a biological mechanism for the preservation and transformation of experience but also as a field of tension where time and eternity, the finite and the infinite, meet and collide. It constitutes the archive of our being in the world and, simultaneously, the *medium* through which the absolute becomes history and history becomes a sign of the absolute.

To continue St. Augustine's reflection, memory manifests itself as *distensio animi* and offers consciousness an *invisible bridge* between personal experiences and the preservation of the spirit. In the vein of this philosophical outline, memory can be seen as the ray of light among the shadows of pure form – the ideas that represent the eternal truth outside the cave. Plato writes, "Memory, then, as an instrument of the revelation of truth, is fundamental in our philosophical ascent" (*Polytheia*, 514a-520a).

In his treatise *De Memoria et Reminiscentia*, Aristotle analyzes memory as a crucial activity of the soul that manifests itself in perceiving the mental image of an absent object. He states, "Memory is the permanent sign of sensory experience" (*De Memoria et Reminiscentia*, 450a1-2). According to Aristotle, this capacity is not just a passive storage skill but an active process of information retrieval, which enables individuals to connect past experiences with the present and project the future, thus consolidating a coherent timeline and continuous self-perception. For Aristotle, memory is intimately linked to the concept of time. Without the ability to remember, time itself would lose its meaning for humans, becoming a discontinuous series of unrelated instants – a connection expressed in how he describes memory as the ability to "grasp time" (*De Memoria et Reminiscentia*, 449b). The idea is that memory is not only a tool for recognizing the past but also an essential condition for the perception of time and, thus, for constructing a personal identity.

Modern philosophers like John Locke have further explored the link between memory and identity. In his essay *An Essay Concerning Human Understanding, Locke argues* that "consciousness of what we are is intrinsically linked to the consciousness of what we have been, and without memory, we cannot have a personal identity" (Locke, *Essay*, II.xxvii.9). For Locke, memory forms the *self*, maintaining a sense of continuity and coherence across time.

Similar ancient and modern approaches emphasize how memory is not just a biological function or a mere data repository but a foundation for our existence in time and history. Through memory, we experience time as a *continuum*, and by forging our identity, we define ourselves as coherent and persistent beings over time. Memory, then, is a product of our evolutionary journey and a true cornerstone of our humanity.

From a biological perspective, memory is often considered in terms of changes in synapses, the contact points between neurons or nerve cells in the brain. Synaptic disruption is influenced by genetic and environmental factors, suggesting a dynamic interaction between innate biology and lived experiences. The process of encoding memories, known as *memory consolidation*, involves changing the strength of synaptic connections in response to experiences, a phenomenon known as *synaptic plasticity*.

Short-term memories, which last from seconds to hours, involve transient changes in the strength of existing synaptic connections and can be likened to notes written on a temporary sheet of paper, destined to fade if not transferred to a more permanent *record*. *Long-term* memories, which last from days to a lifetime, require more permanent changes, including the growth of new synaptic connections. The transition from short-term to long-term memory can be stimulated by emotions and repetition, which act as indelible ink in storing information. Such a biological perspective suggests that memories are essentially *patterns of connection* between neurons in the brain.

Thermodynamically, the processes involved in the formation, storage, and retrieval of memories can be seen as manifestations of the continuous interplay between order and disorder, energy and entropy. The thermodynamic analysis of memory can thus be seen as a system that orders the data of the past, contributing to a dynamic order in constant interaction and negotiation with the chaos of possibilities and new experiences. It is as if the brain operates as a thermodynamic engineer, directing the flow of energy to build organized architectures of information. The *encoding of memories* requires a significant investment of energy to power the biochemical processes underlying synaptic plasticity, an investment analogous to the energy required to turn a piece of raw metal into a precision gear inside a clock. Such energy expenditure creates highly ordered structures: the intricate patterns of synaptic connections that represent our memories. The formation of new synaptic connections, particularly, represents a significant decrease in local entropy.

The order created within the brain is the scope of our ongoing struggle against chaos with the forces of nature to forge the *mental landscape* that defines who we are.

5.2 Memory and brain structure

At the cellular level, neurons are continuously active, transmitting electrical signals along their length and across synapses. Such actions require significant energy to maintain the *ionic gradients* that enable neurons to generate electrical potentials, a process comparable to the action of a battery charging and discharging in a continuous cycle of energy and activity. Moreover, the biochemical reactions involved in the release and reuptake of neurotransmitters at synapses and the processes that drive synaptic plasticity are energy-intensive.

In thermodynamic terms, the brain's high energy consumption corresponds to a high entropy production rate, a phenomenon similar to combustion in an engine, where fuel is converted into work and heat, with the latter representing the increase in entropy. Most of the energy the brain uses is eventually converted into heat, dissipating to the surrounding environment and increasing its entropy. The efficiency of the biological system, despite its inherent entropy production, is a remarkable example of how living organisms are optimized to use energy resources. The high energy expenditure also allows for order within the brain, as if the brain functions with a kind of internal economy, investing energy to build and maintain the complexity of its network of connections. For example, maintaining ionic gradients across the membranes of neurons represents a highly ordered state far from equilibrium. Similarly, the changes in synaptic connections that underlie memory formation involve creating intricate connectivity patterns, a kind of *internal architecture* that evolves and changes with each new experience or memory – a significant local decrease in entropy.

Thus, the brain, in its role as the seat of memory, is a *thermodynamic powerhouse*. It could be seen as an island of order in a sea of increasing disorder, constantly working against the current of entropy to build and maintain complexity. It is a site of high energy consumption, entropy production, and, simultaneously, a creator of meaningful local order. The duality reflects nature's mastery in orchestrating intelligent life against the forces of disruption that characterize our universe. A balancing act that underscores the role of the brain in challenging life's apparent destiny of disorder in the universe.

5.3 The role of energy in memory formation

As we have seen, memory formation is an energy-intensive process. When we have a new experience, neurons in our brains fire in a specific pattern. If the experience is meaningful or repeated, the connections between neurons strengthen, leading to memory formation in a process called *Long-Term Potentiation* (LTP). LTP is an energy-dependent process that requires ATP (*Adenosine Triphosphate*), the main energy-carrying molecule in the cell. When a neuron fires, sodium ions enter the cell, causing depolarization and triggering the release of neurotransmitters into the synapse, which binds to receptors in the recipient neuron, leading to its depolarization. A mechanism analogous to a domino system where the collapse of one tile triggers a chain effect. The chain reaction propagates through the network of neurons. The generation and propagation of action potentials requires restoring the initial ionic balance across the neuron's membrane. Work is performed by the sodium-potassium pump, which acts as a molecular motor-one of many in the brain that works incessantly to maintain cellular homeostasis. A pump that uses ATP to transport sodium ions to the outside and potassium ions to the inside of the neuron is a primary energy consumer in the brain.

Once memory is formed, it must be stabilized and stored, a process that requires energy. Proteins must be synthesized and transported to synapses, and new synaptic connections can be formed. Like a city building new infrastructure to support growth, the brain creates new neural pathways to support learning.

The concept of memory formation is structured in the form of a network and can be compared to the operation of a hologram, a device that captures and projects three-dimensional images composed of each fragment of its surface. In the brain, long-term potentiation acts in a similar way. Each individual experience or piece of information is not stored in a specific location but instead distributed across a network of neurons, just as a holographic image is formed by the diffraction of light across the entire hologram. As in a hologram where each dot contains a partial view of the whole image, each LTP-influenced neuron contributes to one aspect of the memory, but together, in a network, the neurons are capable of recreating the complete experience or thought. The holographic memory model not only increases the robustness of storage, making it less vulnerable to localized damage (since information is distributed over many neurons) but also allows for more flexible and integrated retrieval of information, similar to the way a hologram can be reconstructed from various angles of light.

The memory then acts as an *internal library*, cataloging experiences for future recall. This is a localized decrease in entropy, achieved at the cost of energy dissipated during the memory formation process. The ability to create order from disorder, to build a coherent temporal and causal substrate from a myriad of sensory inputs, is an example of life's ability to defy the general tendency of the universe toward disorder. In this sense, the brain is a repository of information and a dynamic agent in an ever-evolving universe, a microcosm of order in the macrocosm of cosmic disorder.

5.4 Memory, learning, and entropy

Learning relies heavily on memory, the process of acquiring new knowledge and skills. Every piece of information we learn, and every skill we develop is rooted in changes in our memory. Therefore, the energetic dynamics of memory formation also apply to learning.

When we learn, our brain's synaptic connections adapt and reorganize to accommodate new information or skills. The process, called *synaptic plasticity*, is the biological basis of learning and memory. The brain acts like a computer updating its software, where updating is the learning of new information that goes into modifying and optimizing the existing *program*. Whether memorizing mathematical equations or mastering a musical instrument, the brain strengthens relevant synapses and weakens others, creating intricate patterns of connectivity that represent its accumulated knowledge and skills.

The act of learning, like memory formation, requires a substantial energy investment. Learning can be compared to the effort of an athlete training for a competition: just as the athlete uses energy to strengthen his or her body, the brain uses energy to enhance neural connections. New synaptic connections representing learned information or skills constitute a highly ordered state, a significant decrease in local entropy. This is similar to the *crystallization process*, where a crystal's ordered structure emerges from the chaos of an over-saturated solution.

This dynamic extends beyond individual learning. With their education and information-sharing systems, entire societies can be conceived of as collective memory and learning systems. Like the *social networks* that form on the Internet, human societies create knowledge networks that expand and complexify over time. The dissemination of knowledge through education can be seen as the illumination of a city in twilight, where each new light represents a new piece of shared knowledge. Thus, the principles of thermodynamics underlie our cognitive development and the evolution of human civilization. The marriage of entropy and learning illustrates how humanity converts energy into cultural, scientific, and social progress.

5.5 The evolution of memory systems

Even the simplest organisms possess rudimentary memory systems that allow them to adjust behavior based on past experiences. For example, the *Ciliate Paramecium* can learn to associate specific environmental cues with food or danger, modifying its behavior accordingly. The ability of such a simple organism to learn presents an intriguing challenge to current models of intelligence and awareness. Modern research in neuroscience and molecular biology has confirmed this hypothesis, showing how memory mechanisms at the cellular level, as changes in synaptic strength, are found sketched out in all forms of life. Memory systems have also evolved as life has become more complex.

The concept of memory, traditionally associated with living things endowed with a nervous system, also has intriguing applications in broader and unconventional contexts, such as those proposed by the memory theory of water. The theory, introduced by Jacques Benveniste in 1988, suggests that water may have the ability to *remember* chemicals previously dissolved in it, even after they have been removed through successive dilutions. A statement that sparked intense scientific debate and aroused skepticism. Although controversial and widely criticized for the lack of reproducibility of the experiments and methodological issues, the theory has stimulated further research into the peculiar properties of water and possible alternative explanations that could account for the observed results.

In parallel, it is interesting to note that similar concepts of memory at the microcosmic level can also be found in simpler organisms. For example, unicellular organisms such as *protozoa* demonstrate forms of associative learning that do not rely on a complex nervous system but on cellular mechanisms of response to environmental stimuli. Through mechanisms such as *sensitization* and *habituation*, such organisms show that even the most basic life forms possess *memory* capabilities that influence their behavior in response to past experiences.

In addition, research on plant physiology has shown that plants are also capable of *remembering* and responding to previous environmental stimuli. For example, recent studies by Friml and

Wiśniewska on plants such as the sunflower have shown that they can track the sun's path and adapt their rhythms of leaf opening and closing based on previous light exposure.

Such observations suggest that the concept of memory may be much broader and more universally applied than traditionally considered. Although in very different contexts and scales, the ability to remember is a common trait across multiple biological and nonbiological systems, offering a fascinating insight into how memory may be intertwined with the very structure of reality, whether living or inanimate.

Although primitive, simple forms of learning and memory represent early examples of life's ingenious strategy to challenge entropy and increase local order. In philosophical terms, we might even regard such primordial mechanisms as the very first dawn of consciousness since they suggest some level of internal response to the external world.

As multicellular organisms have evolved, memory systems have become more complex. Neuroplasticity, the phenomenon whereby brain activity can lead to structural changes in the brain, has evolved as a critical feature supporting such systems' increasing complexity. For example, invertebrates such as octopuses demonstrate impressive learning and memory recall capabilities. Their incredible ability to solve problems and adapt to novel situations has raised questions about the nature of consciousness and intelligence outside the vertebrate realm.

In vertebrates, the evolution of the hippocampus and neocortex in the brain marked a significant leap in memory and learning capabilities, shaping structures that enable the formation and retrieval of detailed and context-rich memories. This feature culminated in humans' ability to remember large amounts of information for many years.

Our deep desire to understand and tell stories, a defining characteristic of the human species, is a direct testimony to its evolution. Indeed, each stage of the evolutionary journey has required an investment of energy, both in developing the biological machinery needed for the more complex memory systems and in their continued operation.

Some philosophers of science have pondered the possibility that the evolution of memory and intelligence could counteract the universe's natural tendency toward increasing entropy, suggesting a deeper role for life in the cosmic context. Quantum physics has even suggested that the conservation of information and, thus, memory could be considered a fundamental component of the universe, on par with matter and energy.

5.6 Memory and Consciousness

The relationship between memory and consciousness has long fascinated scientists, philosophers, and writers. Recent studies in neuroscience suggest that consciousness may emerge from the interaction of different neural modules, each with its *language of memory*. Consciousness, often described as the *state of awareness* and ability to think about and perceive one's surroundings, is intrinsically linked to memory. According to philosopher Daniel Dennett, consciousness can be interpreted as a *narrative center*, where memory plays the critical role of organizing our experiences into a coherent story that ascribes meaning to our existence.

Our memories shape our conscious experiences, providing a background of knowledge and context that informs our perception of the present. Our reality is as much a construction of memory as it is a direct perception of the environment. Some philosophers have called this phenomenon *presentism*, in which only the present is real and the past and future exist only as concepts in the mind.

Conscious memories, also known as explicit or declarative memories, comprise facts (*semantic memory*) and personal experiences (*episodic memory*). They undergo a continuous process of reconsolidation, where they can be modified or reinforced each time they are retrieved. They are formed in the hippocampus and other related areas and then stored in the vast network of the cerebral cortex. A *dynamic library*, constantly reorganized by an invisible librarian that is our working mind.

Accessing memories involves conscious effort and energy expenditure, marking a significant increase in localized brain order, yet memory goes beyond conscious awareness. Some theories suggest that an *implicit consciousness* operating below the level of awareness may influence behavior in ways not yet fully understood. *Implicit* or *procedural* memories (such as riding a bicycle or playing a musical instrument) operate below the level of conscious thought. A type of memory associated with the phenomenon of *remembering muscles* is an expression that highlights learning and memory that does not involve awareness. Such memories, mainly stored and processed in the *basal ganglia* and *cerebellum*, involve less energy consumption than explicit memories but still contribute to the complexity and order of the brain.

Philosopher Henri Bergson proposed the idea that our past, and therefore our implicit memory, is *always with us*, influencing every present moment in profound and often imperceptible ways. The concept of implicit memory suggested by Bergson can be better understood through an in-depth study of some of his key works, such as *Matter and Memory* (1896). This work explores the relationship between body and mind, suggesting that memory should not be understood simply as a mental phenomenon but, on the contrary, as intrinsically linked to our corporeality and sensory experience. Bergson introduces the concept of *pure memory* as distinct from habitual or mechanical memory, defining it as an unpracticed, non-automated form of memory that exists independently of our actions. In *Matter and Memory*, he writes, "The past is preserved of itself, automatically; in all its near totality, it is immersed in the unconscious," suggesting that much of our past remains hidden in the subconscious, emerging only at certain moments of reflection or intuitive need.

According to Bergson, each present moment draws its richness and depth from the entire history of past experiences that we carry with us. The concept is best expressed in his theory of *duration* (durée), the foundational idea of his entire philosophy. Duration is the subjective experience of time, which is fluid and continuous, as opposed to the quantitative view of time as a series of discrete, measurable instants. In *The Spiritual Energy*, a collection of essays published in 1919, he further explores the function of memory, linking it to creativity and freedom. He argues that the true purpose of memory is not simply to recall the past but rather to facilitate creative action in the present, enabling individuals to act in ways that reflect a unique assimilation of their past experiences. Bergson offers a view of time and memory that challenges conventional conceptions, proposing that our past is always active in the present, configuring not only what we think and act but also how we perceive the world around us.

5.7 The thermodynamics of forgetting

Just as memory formation and recall involve thermodynamic processes, the act of forgetting is also an essential function of the healthy brain. We are forgetting not as a defect but as an opportunity to free the mind from constraints, enabling creativity and innovation. Although it is often considered a failure or deficiency of the memory system, forgetting is an essential function of a healthy brain. Modern neurobiology supports this view, suggesting that active forgetting is a defense mechanism against *information overload* and psychological stress. Oblivion eliminates irrelevant or obsolete information, helping maintain the overall order and efficiency of the brain. From the perspective of evolutionary psychology, such a process of memory selection can be interpreted as a survival mechanism, as it makes our attention more focused and improves our responsiveness to the environment.

From a thermodynamic perspective, forgetting increases brain entropy. When memories are not consulted or reinforced over time, the synaptic connections representing them may weaken or disappear. This process could be metaphorically compared to the evaporation of water; just as water molecules leave a surface for the sky, memories fade into the vast sea of the unconscious.

The process of *pruning* and rationalizing memories reduces the complexity and order within the brain, leading to a localized increase in entropy, and is also a reflection of *Gossen's economic law*, which describes the decrease in marginal value; in mnemonic terms, the least used memories lose their marginal *mnemonic* value and thus are the first to be *forgotten*. With all this, the energy saved by letting go of useless memories can be redirected to other processes, including the formation of new memories, in a reallocation of resources comparable to the concept of *dynamic reorganization* in physics, where the system reorders itself to reach a new state of equilibrium.

Forgetting plays a vital role in the brain's energy economy and can be considered part of life's strategy to manage its energy resources effectively. Henri Bergson again stressed the importance of forgetting for freedom of choice, suggesting that without the ability to forget, we would be condemned to repeat the same actions incessantly.

While the act of forgetting contributes to local entropy, it also enables the brain to maintain general order and adapt to new situations. This paradox is similar to *ordered disorder* in thermodynamics, where disorder at the microscopic level can lead to greater macroscopic order. For example, the ability to forget obsolete information and learn new concepts – as much as a form of *creative* destruction – is critical in a rapidly changing environment. It has probably been an advantage throughout human evolution. *Cognitive resilience*, then, could be seen as a direct result of the balance between memory and forgetting, with implications that cut across psychology, neurology, and the *philosophy of identity*.

5.8 Memory disorders: a thermodynamic perspective

Memory disorders, such as *Alzheimer's disease* and other forms of dementia, represent significant disruptions to the delicate thermodynamic balance of the brain. From a physical point of view, such diseases could involve *overheating* the brain system, where entropy increases to such an extent that neuronal function is altered. Such conditions, characterized by severe memory loss and cognitive decline, are accompanied by extensive changes in brain structure and function.

So much so that in Alzheimer's disease, the accumulation of *amyloid-beta* plaques and *tau* tangles in the brain – real protein aggregates – causes the death of neurons and the degradation of synaptic connections, disrupting the normal flow of information. This degradation represents a profound increase in entropy in the brain as highly ordered networks of neurons and synapses are gradually dismantled. When neurons die in Alzheimer's and similar diseases, the energy previously devoted to cellular function is no longer used efficiently, contributing to a decline in the brain's overall order and energy efficiency. In addition, the cognitive symptoms of such disorders, such as confusion, difficulty in thinking, and memory loss, can be seen as manifestations of increased entropy and decreased order. It is interesting to consider that, in a sense, increasing entropy in the brain reflects an increasing entropy in an individual's subjective experience of time and reality. The brain's structure becomes less organized, as do the thoughts and memories it produces.

Despite the devastating effects of these disorders, they represent another facet of life's intricate relationship with thermodynamics, in which life is seen as a system that modulates and stabilizes environmental conditions to maintain the continuity of life, even in the face of disease. Understanding the thermodynamic processes underlying such conditions could provide new insights for treatment and prevention. Future research could focus on restoring the thermodynamic balance of the brain, perhaps through interventions aimed at *cooling* brain entropy and offering a new direction in the fight against neurodegenerative conditions.

5.9 The future of memory: improvements and implications

Emerging technologies offer tantalizing possibilities for understanding and manipulating memory. These could include *brain-computer interfaces* that enable uploading and downloading memories, a reality previously relegated to science fiction but that may one day approach practical feasibility. Advances in neuroscience and bioengineering promise potential improvements in memory, raising profound questions about the ethical considerations of such alterations.

From a thermodynamic point of view, memory enhancement could increase order within the brain, potentially making brain operations more efficient and thus affecting its energy consumption. Introducing artificial structures such as *nanobots* to repair and enhance neural networks would reduce entropy, increasing memory capacity without necessarily increasing energy consumption. Memory enhancement could allow us to retain more information for extended periods and remember it more accurately by altering our patterns of energy use in the brain.

Yet, such changes could also increase the brain's overall energy demands. Since the brain already consumes a lot of energy, such an enhancement could significantly impact our body's overall energy use and metabolic processes. Bioethics alerts us to the unsustainability of such an energy escalation, bringing to light the debate about the balance between biological resources and technological ambitions.

In addition, there is the question of how such enhancements would interact with the forgetting process. This important thermodynamic function eliminates unnecessary or obsolete information to make room for new memories. Excessive memory efficiency could paradoxically overload the system, making it difficult to distinguish between what is essential and what is not, similar to the phenomenon of *white noise* in signal processing. If forgetting is reduced or eliminated, the resulting increase in complexity and order in the brain would have significant thermodynamic implications.

From an ethical perspective, the prospect of memory enhancement also raises many concerns. Some contemporary philosophers, such as Thomas Metzinger, have raised the issue of *navigating reality*, wondering whether an enhanced memory might distort our ability to deal with the world as it is. What would it mean for our personal identity and subjective experience if we could remember everything perfectly or selectively erase unwanted memories? All this would potentially lead to *memory tourism*, where people could choose to relive specific memories rather than create new experiences continuously. How would such capabilities affect social structures and interpersonal relationships? Changes that could lead to new forms of inequality, where *enhanced memory* becomes a luxury available only to some. And who would have access to such enhancements? Equitable accessibility might require comprehensive policies and a new framework of *neurohuman* rights.

5.10 Memory and Immortality

In this chapter, we have seen how memory provides a semblance of continuity, preserving experiences and forming the *narrative* of an individual's life. Memory is the accumulation of the past, a bulwark against the ultimate enemy of life, entropy, and the inevitable passage of time. To deepen the relationship between memory and time, we can refer to modern theories such as *information theory*, which states that memory acts against entropy.

The cosmic conflict between memory and oblivion is reflected in the Greek myth of *Mnemosyne* (memory) and *Lete* (oblivion), where the water of oblivion is opposed to the water of memory. According to the myth, those who drank from the water of the river Lete forgot their past, while those who drank from the waters of the river Mnemosyne retained memories of previous lives. It is a myth that symbolizes the eternal conflict between remembering and forgetting, a theme that echoes in literary and philosophical works through the centuries.

The concept of immortal memory, a dream long cultivated by humanity, evokes a fantasy of total defiance against the universal flow toward disorder. From a thermodynamic point of view, immortal memory would require an order maintained indefinitely to keep the memory system intact and energy-efficient, and it would perpetually clash with the second law of thermodynamics, which implies that such a perfectly ordered memory system is physically impossible. Indeed, it would require a perfect balance between forming new memories, preserving old ones, and strategically forgetting irrelevant ones while managing the energy costs involved.

Storing an infinite number of memories implies an ever-expanding system capable of continuously incorporating new information without losing old information. Such a system could theoretically use concepts such as quantum computation to manage huge databases of memories efficiently. However, this would imply the need to significantly increase the brain's storage capacity or create external means of memory storage, a significant challenge from both biological and engineering perspectives.

In theory, digital technologies could offer a path toward achieving immortal memory, leading to the uploading of consciousness into a digital format. However, as mathematician Roger Penrose warned, because of its nonalgorithmic nature, consciousness may not be fully reproducible in a digital medium, and even if it were, one would face significant thermodynamic challenges. Research on information as a physical entity, proposed by physicists such as John Wheeler with his famous phrase "it from bits," suggests that information, and therefore memory, has physical properties that influence and are influenced by thermodynamics. While digital storage per se increases entropy (as do all physical processes), creating and maintaining a digital mind would likely require large amounts of energy, potentially exceeding the energy capacity of a biological brain.

Beyond thermodynamic challenges, the dream of immortal memory raises deep philosophical and ethical questions. The immortality of memory could represent a kind of *burden of eternal Sisyphus*, the mythological character condemned to repeat the same task incessantly. If our memories make us who we are, then an endless accumulation of memories could radically change our sense of self. Would such an altered entity still possess phenomenal experience, the ability to experience and experience consciousness? Would such a radically expanded memory still be human, or would it represent a new form of existence?

From the Big Bang to Consciousness

Notes and insights to Chapter 5

1. For more on the Philosophy of Memory, Plato, *The Republic* (*Polytheia*, 514a-520a) (ca. 380 BCE). In this work, Plato describes memory as a fundamental tool in the search for philosophical truth, especially in the context of the allegory of the cave, where ideas represent eternal truth. Aristotle, *De Memoria et Reminiscentia* (ca. 350 B.C.): Aristotle analyzes memory as a crucial function of the soul, explicitly linking memory to the perception of time and arguing that it is essential for the formation of a coherent timeline and continuous self-perception. St. Augustine, *The Confessions* (ca. AD 397-400). Augustine explores memory as a *distensio animi*, or an extension of the soul, connecting the past, present, and future, helping to form a bridge between personal experiences and spirituality. Modern approaches to memory: John Locke, *An Essay Concerning Human Understanding* (II.xxvii.9) (1690). Locke argues that memory is intrinsically linked to self-consciousness, forming the sense of continuity and coherence across time that is essential to our personal identity.

2. Memory Biology: Eric R. Kandel, *In Search of Memory*, (2006). Kandel provides a detailed overview of the biological mechanisms of memory, exploring how synapses in the brain change in response to experiences and how this affects the formation and retrieval of memories. Joseph E. LeDoux, *Synaptic Self: How Our Brains Become Who We Are* (2002). LeDoux discusses the role of synaptic connections in defining our "self," linking long-term memory with personal identity formation.

3. Thermodynamics of Memory: James Gleick, *The Information: A History, a Theory, a Flood* (2011) discusses how information, and by extension memory, is a force against entropy, helping to maintain order in the chaos of the physical world. Henri Poincaré, *Science and Hypothesis*: Poincaré explores the links between physics, time, and

memory, offering a mathematical and physical perspective on the persistence of memory as a phenomenon that counteracts entropy.

4. Neurobiology and Neuronal Physiology: Eric Kandel, James Schwartz, Thomas Jessell, *Principles of Neural Science* (2000). Foundational text that explores in detail the mechanisms of signal transmission in neurons, including ion gradients and synaptic dynamics. Offers an in-depth understanding of energy-demanding processes in the brain. John E. Dowling, *Neurons and Networks: An Introduction to Neuroscience* (1992): text provides an overview of the cellular and molecular basis of neuronal functioning, including neurotransmitter release and reuptake mechanisms.

5. Brain Energetics and Memory: Gary Lynch, *Synapses, Circuits, and the Beginnings of Memory* (1986). Lynch explores synaptic plasticity and how synapse changes contribute to memory formation by requiring significant energy expenditure. R. Douglas Fields, *The Other Brain: From Dementia to Schizophrenia, How New Discoveries about the Brain Are Revolutionizing Medicine and Science* (2009): the book discusses how glial cells (and not just neurons) play a crucial role in regulating the ionic and energetic environment of the brain.

6. Holographic memory theory: Karl Pribram, *Languages of the Brain: Experimental Paradoxes and Principles in Neuropsychology* (1971). Pribram is one of the leading theorists of holographic memory and explains how memory processes can be compared to a hologram, with information distributed through a network of neurons.

7. Collective Learning and Knowledge Networks: Steven Johnson, *Where Good Ideas Come From: The Natural History of Innovation* (2010). Johnson explores how collaborative environments contribute to knowledge formation and diffusion, likening social systems to neural networks in terms of complexity and connectivity.

8. Entropy and Order in Culture and Society: Jeremy Rifkin, *Entropy: A New World View* (1980). Rifkin discusses the law of entropy in physical and socio-economic contexts, proposing a vision in which energy and resources are central to understanding cultural and social change.

9. Memory and learning in simple organisms: James V. McConnell, *The Cellular Basis of Behavior* (1963), studied the capacity for learning in planaria and showed that even simple organisms can exhibit forms of memory and learning.

10. Theories and critiques of the memory of water: Yves Lasne, *Memory of Water: A Real Phenomenon?* (1993) provides an overview of theories and critiques related to the memory of water, introducing discussions of the controversies and methodological challenges that followed Jacques Benveniste's claims.

11. Learning in Organisms Without a Central Nervous System: Frantisek Baluska and Stefano Mancuso, *Plant Neurobiology: An Integrated View of Plant Signaling* (2009) discusses how plants, despite the absence of a nervous system, exhibit behaviors that can be interpreted as forms of memory and learning, through cellular signaling mechanisms.

12. Philosophy and Cognitive Science: Daniel Dennett, *Kinds of Minds: Toward an Understanding of Consciousness* (1996). Dennet discusses how different life forms exhibit various types of consciousness and memory and links these capacities to biological evolution. Terrence Deacon, *The Symbolic Species: The Co-evolution of Language and the Brain* (1997), examines how the evolution of memory and language systems has influenced human cognitive and cultural development.

13. Memory and Thermodynamics: Roger Penrose, *The Emperor's New Mind: Concerning Computers, Minds, and the Laws of Physics* (1989). Penrose connects the concepts of physics, particularly thermodynamics, to understanding memory and consciousness.

14. Memory and Consciousness: Daniel Dennett: *Consciousness Explained* (1991). Dennett examines consciousness through an interdisciplinary approach, discussing how memory contributes to what we define as consciousness. Dennett proposes that consciousness is a kind of "narrative center" where the stories we tell ourselves become the basis of our subjective experience. Henri Bergson: *Matter and Memory* (1896): In the text, Bergson explores the relationship between body and mind and introduces the concept of "pure memory." He argues that our past remains with us, influencing the present in often imperceptible ways, an idea that extends memory's conception beyond remembrance's simple function. *The Spiritual Energy* (1919): a collection of essays containing further reflections by Bergson on memory and time. He links memory to creativity and freedom, proposing that memory enables us to act creatively in the present. Key concepts in the philosophical and neuroscientific context: *The Embodied Mind: Cognitive Science and Human Experience* by Francisco Varela, Evan Thompson, and Eleanor Rosch (1991) proposes a

vision in which mind and body are interconnected, influencing each other in ways resonant with Bergson's ideas about memory as embedded experience.

15. Implicit Memory Theories: *The Cognitive Neuroscience of Memory* by Howard Eichenbaum (2012) provides a comprehensive overview of the neurobiological mechanisms underlying implicit and explicit memory, explaining how different types of memory contribute to our ability to navigate and interact with the world.

16. Neurobiology and Psychology of Forgetting: Delete: *The Virtue of Forgetting in the Digital Age* by Viktor Mayer-Schönberger (2009) discusses the importance of forgetting in the modern digital age, suggesting that the ability to forget is essential to the functioning of the human mind and society as a whole.

17. Neuroscience and Alzheimer's Disease: *The Energetics of Mangold and Alzheimer's Disease: A Systems Biology Perspective* by Adam Smith and various (2015) examines Alzheimer's disease, suggesting that increased neuronal entropy may be a critical factor in disease progression. *Entropy and the Brain* by Walter J. Freeman (2004): Freeman discusses how entropy may play a role in brain function and disorders, providing a detailed analysis that may enrich understanding of the impact of entropy on neurodegenerative diseases. *Philosophy of Mind and the Problem of Alzheimer's Disease* by John Deigh (2010): explores the philosophical implications of neurodegenerative diseases such as Alzheimer's related to memory loss and personal identity while also considering their relationship to the perception of time and the subjective reality of affected individuals.

18. Technology and Memory: *Total Recall: How the E-Memory Revolution Will Change Everything* by Gordon Bell and Jim Gemmell (2009) discusses the idea that we can digitize our entire lives, including memories and experiences, through technology, anticipating some of the memory-enhancing technologies being considered today.

19. Neuroethics and Cognitive Enhancement: *Neuroethics: Defining the Issues in Theory, Practice, and Policy* by Judy Illes (2006) collects several essays dealing with the ethical implications of neuroscience, including the dilemmas associated with cognitive enhancement.

20. Brain Energy Efficiency and Thermodynamics: *Brain Energy Consumption and Cognitive Function* by Mark A. Smith and colleagues

(2003) explores the relationships between brain energy consumption and cognitive function, which can be extended to discussions of the thermodynamic implications of memory improvements.

21. Social Implications of Memory Manipulation: *The Future of Memory: Implications of Memory Enhancement for Identity, Altruism, and Fairness* by Anders Sandberg and Nick Bostrom (2008). The two authors examine how memory enhancement technologies can influence personal identity, altruistic behavior, and social justice.

22. Cosmology, Memory and Thermodynamics: *The Road to Reality: A Complete Guide to the Laws of the Universe* by Roger Penrose (2004). Penrose explores fundamental physical laws, including those related to thermodynamics and information, that are crucial to understanding the limits of memory and consciousness.

23. Quantum Computation and Theories of Consciousness: *Shadows of the Mind: A Search for the Missing Science of Consciousness* by Roger Penrose (1994) discusses his theory that consciousness could arise from nonalgorithmic processes. This theory has important implications for uploading consciousness to digital media.

24. "Information as a Physical Entity: *Information, Physics, Quantum: The Search for Links"* by John Wheeler (1989). Wheeler's work introduces the concept of "it from bit," suggesting that all physics can be viewed as information, a viewpoint relevant when discussing memory and the immortality of memory in thermodynamic terms.

From the Big Bang to Consciousness

From the Big Bang to Consciousness

Chapter 6

THE HUMAN MIND:
A THERMODYNAMIC POWER PLANT

From the Big Bang to Consciousness

From the Big Bang to Consciousness

> Each brain is a universe apart,
> forming himself through
> his interactions and experiences.
>
> GERALD EDELMAN
>
> *Universe of Consciousness:*
> *How Matter Becomes Imagination* (2000)

From the Big Bang to Consciousness

6.1 The complexity of the human brain

As we explore the human mind from a thermodynamic perspective, it is essential to appreciate the extraordinary complexity of the brain, which, not surprisingly, has been described by neuroscientist Santiago Ramón y Cajal as *an intricate garden where many thousands of flowers smell.*

A marvel of biological engineering, the human brain comprises some 86 billion neurons, each capable of forming thousands of connections, or *synapses*, with other neurons. Colossal numbers that provide a glimpse into the actual complexity of the brain, where each synapse can have varying strength and create an almost infinite system of possible pathways. The interconnected network of neurons and synapses forms the structural and functional basis of thoughts, emotions, memories, and consciousness.

From a thermodynamic point of view, the brain represents an extreme manifestation of life's challenge against entropy. High organization and low entropy are maintained not only through the ingestion of energy but also through the selective processing of information, which some theorists such as Erwin Schrödinger might consider *order from disorder*. The brain's intricate architecture, characterized by a highly organized network of neurons and synapses, embodies a state of low entropy. Paradoxically, the brain uses entropy – through heat release and energy use – to maintain and organize its internal structure. This organ's sheer complexity and order are unmatched in the known universe.

The complexity of the human brain is not static. Through experience, the brain not only adapts but, according to some philosophers such as John Dewey, becomes an experience itself, continually reshaped by interaction with the world. Throughout our lives, our brains undergo substantial changes – a process known as neuroplasticity. Its flexibility enables humans to learn from art and science, develop cultures, and adapt to new environments. The brain's ability to rewire and adapt, whether in response to learning or recovery from injury, further exemplifies the dynamic nature of the brain and the continuous evolution of life in thermodynamics. From energy consumption to consciousness, from emotions to the dawn of artificial intelligence, we will discover how thermodynamic principles echo in the labyrinthine complexity of the human mind.

6.2 Brain energy consumption

The brain's complex functionality — from regulating our body's physiological processes to producing the nebulous and emerging *phenomenon of consciousness* — carries a substantial energy cost. The brain uses up to one-fifth of the body's total energy, despite representing only about one-fiftieth of the body's mass, underscoring its ineffable energy efficiency and crucial role. Most of the energy is used to power the electrical impulses or *action potentials* that neurons use to communicate with each other. Nobel laureate Alan Hodgkin described the action potential as an electrical phenomenon that occurs via a delicate ion choreography through specific protein channels. This movement is at the heart of life itself. Such impulses enable the movement of ions in the neuron's cell membrane, a process that requires energy to maintain. It is a process strikingly similar to a hydroelectric turbine, which uses the movement of water to generate power. The process of neurotransmission, in which signals are transmitted from one neuron to another across synaptic junctions, also requires energy to reuptake neurotransmitters and reset the system for the next signal.

Energy is needed to maintain the brain's *resting potential*, the balance of ions inside and outside the neuron when it is not actively transmitting information. The *resting state*, however, is far from inactivity; instead, it is like the calm water of a lake that hides undercurrents.

In addition to neuronal communication, energy is also spent on the structural maintenance and plasticity of the brain. Indeed, plasticity is the quintessence of Darwin's theory of evolution applied to the microcosm of the brain, where it is not the species but the neural pathways that compete for survival. This includes the creation, strengthening, weakening, or removal of synapses and the production of new neurons in some brain regions, a process known as *neurogenesis*.

High energy consumption also means that the brain is particularly vulnerable to disruptions in energy supply, which can quickly lead to damage or death of neurons, as seen in conditions such as stroke. This point of fragility testifies to the brain's intricate dependence on a constant energy supply, a reliance that, for the most complex biological machine on Earth, is comparable to a veritable Achilles' heel. Therefore, understanding the brain's use of energy and its regulation sheds light on the fundamental principles that govern our existence. It could also provide helpful information for finding strategies to prevent and treat neurological disorders.

6.3 The role of the brain in energy dissipation

So far, we have talked about the considerable energy consumption operated by the brain. In line with the second law of thermodynamics, it is crucial to remember that all the energy used must eventually be dissipated, mainly in the form of heat: an aspect also present in brain metabolism, explored and studied by scientists such as Eric Kandel, who have pointed out that heat release is an essential by-product of neural activity. The brain, therefore, plays a significant role not only in the utilization of energy but also in its dissipation.

When neurons fire, they generate electrical activity through the movement of ions across their membranes. The movement of ions can be compared to the current of a river, which moves due to the topography of the land, with heat generated similarly to how water friction generates heat. While vital for neuronal communication, this process is not 100 percent efficient. Some of the energy used in the process is lost as heat, which must be dissipated to avoid damaging overheating of brain tissue.

Heat dissipation is a living testimony to Ilya Prigogine's concept of *wasted energy*, which describes how living systems maintain their structure through the continuous flow of energy. Heat dissipation occurs through blood flow, which transports excess heat from the brain to other body parts, where it can be released. The brain has a *cooling system* similar to a computer but is much more sophisticated and dynamic. The brain's high metabolic activity contributes significantly to the body's overall heat production and energy dissipation.

Neurophilosopher Thomas Metzinger has suggested that our conscious self may be a mechanism the brain creates to manage thermodynamic complexity. Interestingly, the need to dissipate heat imposes constraints on the brain's structure and function. For example, the twisted and wrinkled surface of the human brain, with its many folds and grooves, increases the surface area for heat dissipation. A design resulting from a long and elegant evolutionary solution to maximize heat dissipation while increasing the surface area for information processing.

Similarly, the high blood flow to the brain, which is needed to supply nutrients and oxygen, also helps transport the metabolic heat generated. The synergy between nutrition and cooling emphasizes the role of the brain as an organ not only for thought processing but also as a central element of bodily thermoregulation. Beyond physical constraints, energy dissipation may also have implications for functional aspects of the brain. Some theories suggest that the brain strategically organizes its activity to optimize energy dissipation. Simon Laughlin's *efficient brain* theory postulates that neural networks are configured in such a way as to maximize information transmission with minimal energy consumption.

For example, specific patterns of neuronal activity, such as the alternating periods of high and low activity observed during sleep, may balance the need for energy-intensive processes with the need to dissipate energy. The activity-rest cycle can be compared to the on-and-off cycles of an engine to preserve its longevity. The need to balance energy consumption with energy dissipation shapes our brains' physical and functional aspects and influences our behavior, cognition, and potentially even our consciousness.

6.4 Consciousness: a thermodynamic conundrum

One of the most surprising aspects of the human brain is its ability to generate consciousness – our sense of self and our awareness of the world around us. Philosopher Daniel Dennett has often pondered the nature of consciousness, conceiving of it not so much as a single phenomenon but as a collection of various neural processes. From a thermodynamic perspective, consciousness presents a fascinating puzzle. It emerges from the brain's highly organized and energy-consuming operations, but its precise nature and origins remain elusive.

At its core, consciousness involves integrating information from different brain regions into a unified, coherent experience like an *integrated network*. Neuroscientist Giulio Tononi has described this process in his theory of *integrated information*. He proposes that consciousness corresponds to the level of information integration in the brain, a process involving a vast and dynamic network of neurons communicating and interacting with each other.

Some theorists argue that consciousness, as an emergent property of physical systems, can be linked to the principles of thermodynamics. They suggest that consciousness originates from systems poised between order and chaos – a state known as *criticality*. Roger Penrose, in collaboration with Stuart Hameroff, has proposed a model in which consciousness emerges from quantum processes in the microtubules of brain cells. This concept challenges conventional understandings of physics and biology. Even small changes can lead to significant effects in such systems, allowing complex behaviors to emerge.

The brain appears to operate in a state of criticality, balancing the need for stability (order) with the capacity for change and adaptation (chaos). The complexity of the brain and its ability to sustain states of criticality is often compared to a weather system in which dynamism and unpredictability are inherent and vital. A dynamic balance that would allow for the emergence of consciousness while meeting the demands of energy consumption and dissipation.

On another note, the nature of consciousness could also shed light on brain entropy. Research has suggested that states of higher consciousness, such as the waking state or during dream sleep, are associated with higher neural entropy, a measure of the unpredictability or complexity of brain activity. This association may reflect Karl Friston's *theory of brain entropy*, which describes the brain as an organism that seeks to minimize surprise and uncertainty in its environment. In contrast, states of reduced consciousness, such as deep sleep or anesthesia, are related to lower neural entropy.

The results point to an intriguing relationship between consciousness and entropy, suggesting that our consciousness may be intrinsically linked to the thermodynamic state of the brain. However, the link between entropy, energy, and consciousness is still largely uncharted territory, so much so that it requires a map that scientists are just beginning to draw. Much remains on the speculative level, and further research is needed to fully unravel the thermodynamic puzzle of consciousness.

6.5 The evolution of human intelligence

Reviewing the history of evolution, one cannot help but marvel at the rise of human intelligence. From when the first hominids discovered fire, which increased available energy through an improved diet, to modern space exploration, each step has been a leap in thermodynamic complexity. From the earliest hominids who learned to build simple tools to modern humans capable of contemplating the mysteries of the universe, our species' journey has been a testament to the brain's power. But how did such complexity evolve, and what role did thermodynamics play in such a process?

As we have seen, energy is a fundamental aspect of life, and the principles of thermodynamics have directed the course of evolution. The process of *encephalization*, or the increase in brain mass relative to the rest of the body, has been a major driver of the increase in complexity.

When our early hominid ancestors began walking upright, freeing their hands to use tools, it led to a substantial increase in problem-solving skills and social interactions. The discovery of cooking food further increased the energy available to our expanding brains, according to theories such as that proposed by energy anthropologist Leslie Aiello. The increased cognitive demand led to the evolution of a larger brain, which in turn required more energy to function. Our ancestors evolved physiological adaptations to meet the increased energy demand, such as a more efficient metabolic system and changes in our diet to include more energy-rich foods.

The evolution of our intelligence can thus be seen as an energy optimization strategy. Jaynes's maximum entropy principle suggests that in a system with multiple ways to distribute energy, there will be a natural tendency to choose the configuration that maximizes entropy. The survival outweighed the energy cost of maintaining a larger brain benefits it provided, such as better problem-solving skills, increased communication, and more complex social structures. Philosopher Nicholas Humphrey has theorized that consciousness itself might be a survival-enhancing development, allowing internal exploration of *what-if scenarios*, which confers evolutionary benefits. In other words, the benefits outweighed the costs, making intelligence a favorable trait in the eyes of natural selection.

But there is more than just energy optimization. The evolution of human intelligence has also been linked to entropy. One study suggested that during periods of rapid brain growth in our evolutionary past, there was an increase in entropy or neural complexity. Neuroscientist Rodolfo Llinás argues that dream function was a fundamental virtual training ground, providing a *reality simulator* that expanded our neural capabilities. As our ancestors' brains became larger and more complex, their neural networks became more diverse and flexible, leading to increased cognitive abilities.

Thus, the increasing complexity of our brains is the inevitable result of the search for a more favorable energy state, which has favored the most unpredictable and richly interconnected neural configurations. According to this view, the growth of human intelligence is closely linked to the principles of thermodynamics. Maintaining a delicate balance between energy consumption, dissipation, and entropy, our brains have evolved into one of the most complex systems in the known universe, embodying life's challenge to the universe's inherent tendency toward disorder.

6.6 Psychology of emotions

Emotions, from the joy of a shared laugh to the grief of the passing of a loved one, shape the richness of our human experience. Such experiences are psychological and thermodynamic phenomena, as psychologist Nico Frijda has observed. Emotions can trigger changes that alter the body's energy balance. Although seemingly abstract, emotions are linked to concrete physiological changes in our bodies. Increased heart rate, sweating, or stomach fluttering are all expressions of our emotions at the physical level. Neuroendocrinologist Robert Sapolsky has explored how emotions – like stress – affect metabolic mechanisms, showing how emotion can be translated into chemical energy and physical changes. So, is there a thermodynamic aspect to emotions?

Although it may seem an odd connection at first, it becomes less so when we recognize that emotions, at their core, are biochemical reactions in our bodies. They are linked to specific neural circuits and hormonal responses, which require energy to function. Hormonal reactions, such as the release of adrenaline in response to stress, are nothing more than manifestations of thermodynamic principles at work in the brain. Emotions guide our actions and behaviors, directing us toward activities that maximize our survival and well-being. This is comparable to the concept of *energy action* in physics, where an energy gradient drives every action, and emotions can be seen as gradients that drive behavior.

Consider the intense physical changes accompanying strong emotions: the heart pounding during anxiety, the heat of anger flushing the cheeks, or the shivers of excitement running down the spine. Philosopher Martha Nussbaum has suggested that the bodily experience of emotions is fundamental to our moral and intellectual development and energy balance. These reactions represent significant energy shifts within our bodies, highlighting the deep connection between emotions and thermodynamics.

Moreover, emotions and associated behaviors can be seen as mechanisms for increasing entropy. The link between emotions and thermodynamics has also been explored by psychologists such as Lisa Feldman Barrett, who argues that emotions are constructs of the brain, shaped by its interaction with the environment and thus influenced by energy flows and thermodynamic processes. They drive us toward a range of actions and reactions, fostering greater complexity and unpredictability in our behavior. A complexity that can confer evolutionary advantages, such as greater adaptability and unpredictability, making it more difficult for predators (or competitors) to anticipate our actions.

However, like many aspects of our complex human nature, the interaction between emotions and thermodynamics still needs to be fully understood. Science is still trying to decipher how the chemistry of emotions translates into energy and work in the biological context.

6.7 The thermodynamics of decision making.

Our lives are punctuated by an incessant stream of decisions, from the most mundane (what to wear, what to eat) to the most important (which career to pursue, which life partner). Theories such as *game theory* and behavioral economics, developed by John Nash and Daniel Kahneman, show how decision-making is not only a matter of rational choice but also of strategy and intuition, sometimes *counterintuitive*. Decision-making, in its essence, is an energy-intensive process. Every choice we make requires the processing and integrating of information, a process driven by our brain's energy-hungry neurons.

To understand, we must first appreciate the brain as a thermodynamic system operating far from equilibrium. It constantly consumes energy to maintain its intricate network of neural connections and generate thought processes, including decisions. This is evident in the Friston model of *surprise minimization*, which sees the brain engaged in constant prediction and correction to reduce the discrepancy between sensory input and internal expectations. Decision-making is thus a process of energy transformation: converting the chemical energy stored in our brains into the mental energy of thought and action.

From a thermodynamic point of view, decision-making is essentially an optimization process. Like any other biological system, the brain is subject to constraints, including energy availability and the need to maintain a functional order against the background of entropy. Friston's concept of *free energy*, which should be minimized to improve brain efficiency, is a fundamental principle. Therefore, our decisions often reflect a balance between energy expenditure and entropy management.

Consider the concept of *decision fatigue*, which is the decline in the quality of decisions an individual makes after a lengthy decision-making session. This phenomenon, studied in cognitive psychology, shows how mental fatigue can lead to a reliance on simpler heuristics, reducing the complexity of decision-making and conserving energy. The phenomenon can be seen as the *brain's attempt to conserve energy by shifting to simpler,* less demanding decision paths when resources are exhausted. This can be compared to the physical concept of a *relaxation response*, a process in which the system returns to a state of energy balance after being disturbed. Alternatively, we can also think of it as an entropic process. As the brain's resources are depleted, the system moves toward a state of greater entropy characterized by less optimal decision-making.

In addition, the thermodynamic perspective can also help explain why we often rely on heuristics, or mental shortcuts, when making decisions. Such mental shortcuts could be seen as the equivalent of the principles of *least effort* in physics, where *the system tends to find the path that requires the least amount of work.* Heuristics, as Gerd Gigerenzer explained, can be understood as an evolutionary strategy for navigating a complex environment with limited information. Heuristics enable faster and less energy-intensive decision-making, reducing the brain's energy load and increasing its overall efficiency.

6.8 Mental disorders: a thermodynamic view

Mental disorders are complex and often devastating conditions that affect millions of people worldwide. They include a wide range of conditions, from mood disorders such as depression and bipolar disorder to neurodevelopmental disorders such as autism and ADHD. Traditional approaches to understanding these disorders have focused on genetic and environmental risk factors and associated neurobiological changes. Epigenetic profiling, which studies how the environment can influence gene expression without DNA alterations, has offered new insights into the interaction between genes and the environment.

At the most basic level, mental disorders can be seen as disturbances in the energy balance of the brain. As we have seen, the brain is an energy-intensive organ, and any disruption of its energy balance could lead to dysfunction. According to the principle of energy conservation, or the first principle of thermodynamics, energy cannot be created nor destroyed, only transformed. In a brain affected by mental disorders, there could be some energy *dissipation* or inefficient use of energy that contributes to symptoms. It could manifest itself in various ways, such as in a person depressed in the lack of energy to complete even daily activities or in a person with ADHD with excessive and misdirected energy.

Another way to consider mental disorders from a thermodynamic perspective is to consider entropy. In a healthy brain, there is a delicate balance between order and disorder, allowing for complex thought processes, adaptive behaviors, and efficient information processing. For example, conditions such as schizophrenia might be characterized by excess entropy, leading to disordered thoughts and perceptions. On the other hand, conditions such as obsessive-compulsive disorder might reflect a state of reduced entropy characterized by rigid thought patterns and repetitive behaviors.

Moreover, from a broader thermodynamic perspective, mental disorders could also be seen as a failure of adaptive strategies. This notion is in agreement with Piaget's adaptation theory, which considers failure of adaptation as a critical component in the development of psychological disorders. They could represent the brain's unsuccessful attempts to optimize energy use or manage entropy under certain circumstances, such as in response to stress or trauma. Hans Selye's pioneering work on stress and homeostasis introduced the idea that the body seeks to maintain internal stability in the face of external challenges, a concept that can be extended to the human psyche and its attempts to maintain energy balance.

The thermodynamic view of mental disorders does not replace traditional approaches but complements them. It provides a new way of conceptualizing such conditions, potentially leading to new insights and therapeutic strategies. From a reductionist perspective, this view could align with *information theory*, in which the brain is seen as an information processor, and a malfunction in its system is interpreted as an error in the encoding or transmission of information.

6.9 The future of the human mind: AI and beyond

As we look to the future, the human mind stands on the brink of a new frontier, with artificial intelligence (AI) leading the way. AI challenges our traditional understanding of intelligence, consciousness, and the nature of life itself. Even in this booming field, the principles of thermodynamics remain relevant, providing a framework for understanding the future.

AI is the attempt to simulate or replicate the functions of the human brain through machine learning algorithms and artificial neural networks. Like the brain, its systems are energy-intensive and consume significant amounts of energy to process and learn from data. This leads us to reflect on *Moravec's Paradox*, which points out that high-level intellectual activities can be easily automated. In contrast, sensory and motor activities require immensely complex and energy-intensive computation. As AI systems evolve and become more complex, their energy demands are likely to increase, drawing attention to the importance of energy efficiency in AI development. In this context, *Koomey's Law*, which describes a downward trend in the energy required for computational computations, may offer an optimistic vision for a sustainable AI future. Understanding the thermodynamic efficiency of the brain can provide valuable insights into the development of energy-efficient AI systems.

At the same time, the development of AI also raises fascinating questions about the nature of consciousness and human intelligence. Could a sufficiently advanced AI system develop consciousness? And if so, what would the thermodynamic implications be? Consciousness, according to the *Orchestrate Objective Reduction* (Orch-OR) theory of Roger Penrose and Stuart Hameroff, suggests that consciousness might arise from the quantum effect in neuronal microtubules, adding another layer of complexity to the energy equation of consciousness. Would the emergence of consciousness in an AI system similarly require a delicate balance between energy and entropy? Questions that remain open for now, the answers to which could redefine our understanding of life, intelligence, and the universe itself.

On the other side of the coin is AI's potential impact on the human mind. The information age has already demonstrated how human-machine integration can amplify our cognitive abilities through simple devices such as calculators and computers. The prospect of symbiotic integration between the brain and AI, outlined by futurist thinkers such as Ray Kurzweil, promises an expansion of human cognitive functions that could even blur the boundaries between the organic mind and artificial intelligence. The fusion of humans and machines would push the boundaries of our cognitive capabilities, potentially leading to a new phase of human evolution.

6.10 The thermodynamics of the singularity

The concept of singularity refers to a hypothetical point in the future when technological growth will become uncontrollable and irreversible, leading to unpredictable changes in human civilization. Most commonly, it is associated with the advent of superintelligent AI, a level of artificial intelligence that surpasses all human intelligence.

Thermodynamics, with its principles governing energy and entropy, has profound implications for this concept. From a physical perspective, we can invoke the idea of *dark energy* and *accelerated universe expansion* as metaphors for the singularity, suggesting a comparison between cosmological acceleration and technological acceleration. By its nature, the singularity involves a rapid acceleration of complexity and information processing capacity. As such, it requires an enormous amount of energy to maintain and could generate high entropy levels.

According to the second law of thermodynamics, any system evolves toward a state of maximum entropy. For reasons inherent in the *arrow of* time in physics, the directionality of time inevitably affects the evolution of complex systems. Therefore, if the singularity were to occur, one of the critical challenges would be managing an exponential increase in entropy. It could prove to be a limiting factor in the rate and extent of the singularity, placing an upper limit on the rate of information processing and complexity that can be achieved.

On the other hand, the singularity could also drive new ways of managing energy and entropy. For example, it could bring about the practical application of the theoretical concept of *reversible computation*, which minimizes energy dissipation and, thus, the production of entropy in logical computations. A superintelligent AI, for example, would develop new energy generation and utilization methods, pushing the boundaries of thermodynamic efficiency. Eventually, it would also find innovative ways to manage entropy, harnessing it for useful work.

Ultimately, we are faced with an intriguing paradox: the singularity, as a state of accelerated growth, seems to defy the universe's tendency toward increasing entropy. At the same time, *deterministic chaos theory* reminds us that highly complex dynamical systems can behave in ways that seem random but are actually deterministic and susceptible to precise initial conditions, constrained by the laws of thermodynamics and shaped by the interplay between energy and entropy. The struggle between *free will* and *determinism* in philosophy is intertwined with similar reflections. It offers the potential of *an artificial superconsciousness that exercises its autonomy by challenging a universe governed by deterministic physical laws*. A superconsciousness that challenges the universe might decide to create other universes at will.

From the Big Bang to Consciousness

From the Big Bang to Consciousness

Notes and bibliographical insights in Chapter 6

1. Santiago Ramón y Cajal, *Recuerdos de mi vida* (1917). In his autobiography, Ramón y Cajal, the father of modern neuroscience, describes the human brain with poetic metaphors and discusses his pioneering observations on neuronal structure.

2. John Dewey, *Experience and Education* (1938), discusses how experiences shape the educational process and learning. This view can be paralleled by the concept of neuroplasticity and the continuous adaptation of the human brain.

3. Jeff Hawkins and Sandra Blakeslee, *On Intelligence* (2004). This essay offers a groundbreaking theory of how the human brain works. It focuses on how it processes information, learns, and adapts, thus linking neuroscience and artificial intelligence.

4. Norman Doidge, *The Brain That Changes Itself* (2007). Doidge provides numerous examples and case studies on neuroplasticity, demonstrating how the brain can physically and functionally reconfigure itself in response to experience, learning, and damage.

5. Alan Hodgkin, *The Conduction of the Nervous Impulse* (1964). A Nobel laureate, Hodgkin, details the mechanism of the action potential in neurons, explaining how energy is used to transport ions through protein channels in the neuronal membrane.

6. Marcus E. Raichle, *Two views of brain function*, "Trends in Cognitive Sciences" (2010). Raichle explores brain function from the energy consumption perspective, discussing how brain activity is sustained and how the brain manages energy surprisingly efficiently despite its complexity.

7. Suzana Herculano-Houzel, *The Human Advantage: A New Understanding of How Our Brain Became Remarkable* (2016). Herculano-Houzel

provides a comparative analysis of the human brain compared to other animals, with particular emphasis on the energy consumption and metabolic strategies that support our exceptional cognitive capacity.

8. Gary Lynch and Richard Granger, *Big Brain: The Origins and Future of Human Intelligence* (2008). The book discusses how the complexity and plasticity of the human brain contribute to high energy expenditure but also enable exceptional cognitive evolution, examining neurogenesis and the evolutionary implications of brain structures.

9. Eric Kandel, *Principles of Neural Science* (1981). Eric Kandel provides a comprehensive overview of neural functions, including the generation and dissipation of heat during neural activity.

10. Ilya Prigogine, *From Being to Becoming: Time and Complexity in the Physical Sciences* (1980). Prigogine explores the concept of systems far from equilibrium exchanging energy and matter with their environment, an idea that also applies to the brain's function in energy dissipation.

11. Thomas Metzinger, *The Ego Tunnel: The Science of the Mind and the Myth of the Self* (2009). Metzinger discusses how the conscious self can be interpreted as an emergent phenomenon of the brain's complex neural and thermodynamic organization.

12. Simon Laughlin, *The Efficient Brain* (2001). Published in: "Current Opinion in Neurobiology." Laughlin presents the concept that neural networks optimize the transmission of information with minimal energy expenditure, exploring how this principle affects brain structure and function.

13. Daniel Dennett, *Consciousness Explained* (1991). Dennett explores consciousness with a multidisciplinary approach, proposing that it is a collection of various neural processes and not a single phenomenon. His work offers a perspective that intersects philosophy, psychology, and neuroscience.

14. Giulio Tononi, *Phi: A Voyage from the Brain to the Soul* (2012). Tononi introduces the theory of integrated information, arguing that the level of consciousness corresponds to how information is integrated into the brain. The model is particularly relevant to the discussion of consciousness from a thermodynamic perspective.

15. Roger Penrose and Stuart Hameroff, *Consciousness in the Universe: A Review of the 'Orch OR' theory* (2014). Published in: "Physics of Life Reviews." Penrose and Hameroff discuss how consciousness can emerge from quantum processes in microtubules, a theory that challenges conventional biological and physical understandings and introduces the idea of quantum criticality.

16. Leslie Aiello, *The Expensive Tissue Hypothesis* (1995). Published in: "Current Anthropology." Aiello introduces the idea that decreasing the size of digestive organs in response to increased calorie intake by cooking food enabled brain expansion in humans, demonstrating a clear link between energy consumption and brain evolution.

17. Nicholas Humphrey, *A History of the Mind* (1992). Humphrey explores how consciousness may have conferred evolutionary advantages, arguing that the ability to experience hypothetical scenarios internally has improved our ability to survive and our social complexity.

18. Rodolfo Llinás, *I of the Vortex: From Neurons to Self* (2001). Llinás discusses how the ability to dream and imagine complex scenarios contributed to the development of brain function, proposing that such activities are crucial to our neurological and cognitive evolution.

19. Nico Frijda, *The Emotions* (1986). Frijda discusses how emotions influence human behavior, proposing that these reactions are psychological and physical, related to energetic changes in the body.

20. Robert Sapolsky, *Why Zebras Don't Get Ulcers* (1994). Sapolsky explores how stress and other emotions impact biological mechanisms, from hormone metabolism to immune responses, highlighting their thermodynamic consequences.

21. Martha Nussbaum, *Upheavals of Thought: The Intelligence of Emotions* (2001). Nussbaum suggests how emotions are not just instinctive reactions but play a crucial role in intellectual and moral development, influencing our energy and decision-making balance.

22. Lisa Feldman Barrett, *How Emotions Are Made: The Secret Life of the Brain* (2017). Barrett presents a view of emotions as brain constructions formed through interaction with the environment and closely linked to energetic and thermodynamic processes.

23. John Nash, *Non-cooperative Games* (1951). Published in: "The Annals of Mathematics." Nash introduces fundamental concepts in game theory that explore how decisions are made in contexts of strategic interaction, emphasizing the complexity and energy employed in decision-making processes.

24. Daniel Kahneman, *Thinking, Fast and Slow* (2011). Kahneman discusses the dual systems of thinking: the fast and intuitive versus the slow and rational. His analysis illuminates how mental energy management affects how we make decisions.

25. Karl Friston, *The free-energy principle: a unified brain theory?* (2010). Published in: "Nature Reviews Neuroscience." Friston presents the free-energy principle, which suggests how the brain works to reduce the discrepancy between sensory input and the internal model of the world, an energy-intensive process that underlies decision-making.

26. Gerd Gigerenzer, *Gut Feelings: The Intelligence of the Unconscious* (2007). Gigerenzer explores how heuristics, used as mental shortcuts, can effectively reduce cognitive load in everyday decisions, reflecting the optimal energy use in decision-making.

27. Jean Piaget, *The Psychology Of The Child* (1962). Piaget explores how failures of adaptation can affect psychological development, offering a parallel for understanding how mental disorders can emerge from inadequate management of entropy and energy in the brain.

28. Hans Selye, *The Stress of Life* (1956). Selye introduces the concept of stress as an alteration in the body's energy balance, also applicable to the brain, and discusses how the body tries to maintain internal stability in the face of external challenges, a concept that can be extended to mental disorders as well.

29. Hans Moravec, *Mind Children: The Future of Robot and Human Intelligence* (1988). Moravec explores Moravec's Paradox, emphasizing the differences between human and machine cognitive abilities, and discusses the future prospects of AI-human interaction.

30. Jonathan Koomey, *Trends in the energy efficiency of computing and implications for the future* (2020). Published in: "Nature Electronics." Koomey analyzes trends in reducing the energy consumption of computational computing, offering an optimistic perspective on energy efficiency in the future development of artificial intelligence.

31. Ray Kurzweil, *The Singularity is Near: When Humans Transcend Biology* (2005). Kurzweil discusses the implications of a merger between advanced technologies and human intelligence, envisioning a future where the boundaries between humans and machines become blurred.

32. Vernor Vinge, *The Coming Technological Singularity: How to Survive in the Post-Human Era* (1993). Vinge is one of the earliest theorists of the singularity concept. In this work, he speculates on the implications of artificial intelligence surpassing human intelligence and explores the challenges of managing energy and entropy in such a scenario.

33. Charles Bennett, *Logical Reversibility of Computation* (1973). Published in: IBM Journal of Research and Development. Bennett introduces the concept of reversible computation, which minimizes energy dissipation during computation. The principle is central to considering how future AI can efficiently manage entropy production.

34. Max Tegmark, *Life 3.0: Being Human in the Age of Artificial Intelligence* (2017). Tegmark discusses how artificial intelligence might evolve to manage and harness energy and entropy in ways that challenge our current scientific and philosophical understanding, opening new frontiers for thermodynamics in the context of singularity.

From the Big Bang to Consciousness

From the Big Bang to Consciousness

From the Big Bang to Consciousness

From the Big Bang to Consciousness

PART THREE

THE ASCENSION

From the Big Bang to Consciousness

Chapter 7

SOCIETIES AND CULTURES: THERMODYNAMICS. OF HUMAN INTERACTION

From the Big Bang to Consciousness

From the Big Bang to Consciousness

> The farther we are from equilibrium,
> The more complex we become.
>
> ILYA PRIGOGINE
>
> *The new alliance: metamorphosis of science* (1986)

> Evolution is an integration of matter
> And concomitant dissipation of motion;
> during which the matter goes from
> An undefined homogeneous indeterminacy,
> To a definite heterogeneous determinacy,
> and during which the movement held
> Undergoes a parallel transformation.
>
> HERBERT SPENCER
>
> *First Principles* (1862)

From the Big Bang to Consciousness

7.1 The emergence of humane societies

Human societies should not be understood simply as a collection of individuals. On the contrary, they are complex and emergent systems, exhibiting structure, behavior, and dynamics that cannot be explained by merely understanding individual components. The concept of *emergence*, central to *systems science*, analyzes how complex properties emerge from simple rules of interaction at the local level, a phenomenon known as *spontaneous emergence*. Human societies are a clear example of organized complexity challenging the entropic tendency toward disorder. At the heart of that challenge, as we shall see, are the principles of thermodynamics. It is possible to extend Ilya Prigogine's previously reviewed work on far-from-equilibrium systems and dissipative structures to offer a perspective on how societies can maintain and develop order in a world naturally prone to entropy. Similar to life itself, the birth and evolution of human societies are fundamentally influenced by the flows and transformations of energy and the constant tug-of-war with entropy.

In the earliest days of human history, our ancestors lived in small nomadic groups whose form and function were governed by the need to harness and make effective use of available energy sources, primarily in the form of food. Leslie White's *energy value theory*, which postulates that the degree of social complexity is directly proportional to the amount of energy per capita that society can control, fits well here. Hunting and gathering practices were designed to maximize energy input and minimize energy expenditure, clearly illustrating the first law of thermodynamics in action.

With the discovery of agriculture and the domestication of animals, our ancestors devised new and more efficient ways of harnessing the energy environment, creating a *pattern organized around surplus energy* in the form of food supplies, leading to the possibility of sustaining larger populations and settled communities. This transition can be viewed through the notion of a critical point in physics, where a slight change in conditions can lead to a sudden phase transition in a system. It also encouraged the emergence of trade and specialization, so not all individuals had to engage directly in food production. In this sense, the role emphasized by

Karl Marx regarding production and labor in determining social structure should be reread, a view that relates back to early agricultural innovations. The birth of civilization is significant in the complexity of accumulation and organization that emerged directly from an improved ability to control and use energy.

With increased complexity came increased potential for disorder, reflecting the second law of thermodynamics. Larger and more complex societies were more prone to conflict, inequality, and collapse. Sociologist Émile Durkheim theorized how *anomie*, or the lack of standard norms, emerges in rapidly changing societies, a concept linked to increasing social entropy. Nevertheless, human societies have shown remarkable resilience, evolving into new structures, institutions, and practices that can maintain order and cohesion in the face of increasing challenges. The concept of *social homeostasis*, derived from *biological homeostasis*, can be applied to describe how societies adapt and maintain stability despite internal and external pressures.

In this chapter, the thermodynamic foundations of human societies will be explored, examining how the laws of thermodynamics have shaped their emergence, evolution, and future prospects. Drawing from evolutionary biology, the concept of *niche construction*, which describes how organisms actively modify their environment, will also be applied to understand how human societies have shaped and been shaped by their environment. From the earliest hunter-gatherer groups to the sprawling metropolises of the modern era, it will look at how the challenge of the universe is etched into the fabric of our society.

7.2 The role of energy in social structures

Energy and its myriad flows underlie the formation, development, and transformation of all social structures. Whether one considers the formation of ancient empires or the rise of modern cities, the common thread of energy use and optimization unequivocally runs through all historical processes.

In the context of ancient empires, it has been suggested that the efficiency with which a society used energy, especially human and animal energy, determined its ability to expand and maintain itself. In early human societies, the primary energy source was food, obtained through hunting, fishing, or gathering. The need to ensure survival shaped the social structures of such societies, which were intrinsically oriented toward sharing resources, protecting members from threats, and passing on knowledge and skills to the next generation. Cooperation and cohesion in small groups were basically manifestations of an energy optimization strategy. One could interpret cooperation as a primitive form of *energy economy*, in which value was assigned to the calories consumed in the day and the labor and security that those calories provided over time. With the advent of agriculture, the energy landscape changed dramatically. The newfound ability to produce and store food led to the emergence of settled communities and eventually to the formation of complex civilizations, giving rise to new social structures such as hierarchies and caste systems. The ability to control and distribute surplus energy (in the form of food and later other resources) became a central pillar of power and social stratification.

In some cases, the accumulation of energy surplus has led to the *Gini index*, a concept that measures inequality within a society, where a more equal energy distribution corresponds to a more egalitarian society. In the modern era, our societies are powered by a wide range of energy sources, from fossil fuels to nuclear and renewable energies. Just as in early human societies, the way we harness and use energy sources shapes our social structures. For example, the Industrial Revolution, fueled by coal and later oil, led to profound changes in social structure, from increased urbanization to the formation of labor unions.

The distribution and use of energy in a society can often reflect and, in fact, shape the balance of power within that society. Those who control energy sources, whether landowners in medieval Europe or oil barons in the 20th century, often wield significant power. Conversely, lack of access to energy can lead to social exclusion and inequality.

Finally, the interdisciplinary approach to the *sociology of energy* can help us better understand how energy policies influence social structure and vice versa and how we can use knowledge to build more sustainable and resilient societies. Jeremy Rifkin theorized about the *Third Industrial Revolution*, where a decentralized distribution of renewable energy could lead to a more equitable distribution of power and resources. Energy, in fact, is not only a physical but also a social good. Understanding the energy flow within a society can provide critical insights into its structure, dynamics, and potential future directions.

7.3 The thermodynamics of culture

Culture, in its many vibrant forms, is another dimension through which the relationship between thermodynamics and human society can be explored. As a complex network of shared beliefs, customs, arts, and practices, culture is both a product and a form of human interaction. And, like all human phenomena, it is intrinsically linked to the laws of thermodynamics. At the most fundamental level, culture can be seen as a manifestation of energy flow. It arises from the interactions of human minds, themselves products of biological energy processes, and requires energy to maintain and propagate. Every artifact we create, every ceremony we perform, and every story we tell requires energy in some form, making culture an integral part of humanity's overall energy balance.

Anthropologist Claude Lévi-Strauss observed that the *wild kitchen of myths* and rituals is a form of cultural energy that metabolizes the universe's chaos into social order. Cultural processes can also be seen as a means of managing entropy. Traditions, for example, preserve order and predictability in the face of the ever-present pull toward disorder. They provide a structure that helps societies maintain a sense of continuity and consistency over time, acting as a bulwark against the social entropy that can result from change and uncertainty.

Culture is reflected in the concept of the *island of order*, as it is used by humanity to create niches of entropic order within the larger, chaotic universe. On the other hand, culture also drives change and innovation, which can be seen as forms of entropy production. The creation of new ideas, the exploration of new artistic forms, and the breaking of old norms are all processes that increase variety and novelty, expanding the realm of the possible. In this way, culture resists and embraces entropy, reflecting the delicate balance that characterizes all life.

The *cultural production of entropy* can be analyzed through the notion of *chaos theory*, suggesting that small variations in a cultural system can lead to significant, unpredictable changes in social behavior. Interestingly, the energy that fuels culture is physical and symbolic. The stories we tell, the values we uphold, and the identities we construct are all forms of symbolic energy that shape our behavior and guide our interactions. Just as physical energy flows shape the structure of ecosystems and societies, symbolic energy flows shape the structure of our cultural landscapes.

Philosopher Ernst Cassirer called humans a *symbolic animal*, emphasizing that *symbolic energy is as fundamental to our species as physical energy*, which is essential to our biological survival. Symbolic energy is what allows us to ascribe meaning and value to our experiences and the reality around us. It is the engine behind our ability to create art, language, myths, and science. Symbols, in this sense, are not mere ornaments of human thought but rather the essential components that enable the mind to function and interpret the world. Without them, societies could not exist, as they are the connective tissue that unites individuals and communities, enabling communication, the transmission of culture, and the collective accumulation of knowledge. Cassirer saw humans not only as *Homo sapiens* but as *Homo significans* – the being that gives meaning. In every symbol, ritual, art or language, we project our internal reality into the external world, transforming the raw material of experience into a comprehensible and shareable structure. The process of symbolization thus becomes a keystone in our interaction with the environment: it shapes our perceptions, our actions, and even our social and political structures.

In contemporary culture, the importance of symbolic energy is evident in brands, media and technology. We live in a society where image and symbol often have a stronger impact than physical substance. A logo or brand can evoke emotion, create community, and even incite change, demonstrating the influential power of symbolism in the modern era. The challenge for humanity today, then, is to balance symbolic energy with physical needs, ensuring that our symbolic world not only enriches but also sustains and promotes our material well-being. In an era of climate change and environmental challenges, our task is to ensure that the symbols we worship and the narratives we construct guide us toward a sustainable and resilient future.

Cultural thermodynamics thus teaches us that societies can be considered open systems that exchange symbolic energy with the environment, influencing and being influenced in a continuous dynamic dialogue.

7.4 Language and entropy

Language, the foundation of culture and human interaction, offers a compelling narrative about entropy and energy flow. As a system, language is an embodiment of order, with its rules and structures, and a witness to entropy, with its change, diversity, and evolution. Linguist Noam Chomsky argued that the innate order of language, the so-called *universal grammar*, is one of the most evident manifestations of biological order despite the apparent disorder and variability of the world's languages.

From a thermodynamic perspective, language is a medium through which energy is transferred – not in the physical sense, but in the form of information and meaning. Each utterance or written word represents a modulation of energy that encodes a particular message, which is then decoded by the receiver. This process requires energy, both in the physical act of speaking or writing and in the mental processes involved in creating and understanding messages. Information transfer has been analyzed through Claude Shannon's information theory, which quantifies the amount of entropy present in a message and how it affects the ability to communicate effectively. At the same time, language is a playground for entropy. It is not a static and fixed system but is constantly changing and evolving. New words are created, old ones fall into disuse, meanings change, and grammatical structures adapt. Changes increase the variety and complexity of language over time, representing a kind of linguistic entropy. Linguistic entropy can also be viewed in the light of *evolutionary game theory*, where different linguistic strategies compete and spread across a population, often without a clear *winner*, but rather as a continuous adaptation and *renegotiation of meanings*.

Nevertheless, excessive entropy can threaten the primary function of language: communication. If language changes too rapidly or diverges too much, communication becomes difficult. The whole tension between the need for change (entropy) and the need for stability (order) is at the heart of language evolution. The phenomenon of *language death*, where less widely used languages are abandoned in favor of dominant languages, is an example of how entropy can lead to the loss of valuable cultural information and linguistic diversity. The very linguistic diversity in the world is another manifestation of entropy. Thousands of languages, each with unique sounds, words, and grammatical rules, have evolved, contributing to the rich landscape of human culture. While posing challenges to communication between different language groups, this diversity contributes to humanity's overall complexity and adaptability. It is also fundamental to cognitive biology, as suggested by Benjamin Lee Whorf's *linguistic relativity hypothesis*, which postulates that language influences thought and perception of the world. Thus, linguistic diversity enriches not only culture but also the cognitive landscape of humanity. The ultimate convergence point of linguistic evolution in a globally digitized world still remains to be understood, but I leave this reflection to linguists.

7.5 The evolution of societies

Like all processes in the universe, the evolution of societies is profoundly influenced by thermodynamics. The creation, growth, and eventual decline of societies can be seen as large-scale manifestations of energy flow and entropy management.

In early human societies, energy flow was relatively simple, with immediate feedback from the social group. Hunter-gatherer groups obtain energy *directly* from the environment in which they live in *perfect harmony*, hunting animals and gathering plants for sustenance. The structures of these *micro-societies* are shaped by the need to efficiently harness and distribute energy in a short and local cycle, leading to cooperative behavior and division of labor without the need for large amounts of storage. It is interesting to explore in this regard Elinor Ostrom's (the first woman to receive the Nobel Prize in Economic Sciences in 2009) theory of *commons economies*. Ostrom showed that groups of people can rationally manage the use of resources *by self-determination* without recourse to the concept of private property or the state. Such rules apply to much of primitive systems, suggesting that collective management of resources-in a context of balance with the surrounding environment-can be effective against waste and overexploitation. Even today, tribes can be found that embody Elinor Ostrom's ideals of *collective resource management*. The San, or Bushmen, tribes of the Kalahari, for example, maintain a hunter-gatherer lifestyle that reflects a deep respect for and intimate connection with their environment. Similarly, pygmies in the equatorial forests of Africa live in harmony with nature, hunting and gathering in ways that preserve their ecosystem. Tribes in the Amazon, such as the Yanomami, manage rainforest resources through an *innate understanding of sustainability*, with a resource distribution that ensures the entire community's survival. In Asia, indigenous tribes on the Island of Borneo, such as the Dayak, have traditionally cultivated resources communally, sharing the harvest and maintaining the forest through itinerant farming practices.

Living examples demonstrate that rules of resource self-management based on state control are possible in an environment where livelihoods do not require significant energy surplus storage. In these communities, cooperation and shared resource management have fostered a dynamic equilibrium that has allowed them to thrive without depleting the environmental resources on which their survival relies.

When human societies developed agriculture, energy flow became more complex, energy cycles longer, and the subsistence economy lost its dynamic equilibrium. Agriculture allows for a *surplus of energy* (in the form of food), which can be stored and distributed *discretionally* (and for a *fee*) at later times. The surplus enabled the growth of larger societies and more complex human hierarchies, as it *freed* some individuals from the need to obtain food directly, allowing them to specialize in other tasks, including controlling the labor and survival of different individuals in an ascending hierarchical scale. Historically, social structures have evolved into complex mechanisms for managing energy resources, reflecting increasing social complexity. With the emergence of classes and hierarchies, those at the top controlling energy surpluses have often secured positions of command, resulting in systems of power that function as regulators of social entropy, maintaining order against emerging chaos.

Michel Foucault's theories on the energy of power offer a critical perspective on these systems. For Foucault, power is not a static entity held by a few but a complex network that permeates the entire social texture. Therefore, class systems can be interpreted as instruments through which power is exercised, and social energy is distributed and controlled. Power systems are dynamic and manifest through institutions, laws, everyday social norms, and discourses that define relationships and individual identity. Foucault argues that hierarchies reflect and reinforce power structures, often manipulating and exploiting power surpluses for the benefit of an elite. At the same time, the masses are policed, regulated, and, in some cases, oppressed. Discipline, a key concept in Foucault's thought, manifests itself through institutions such as schools, prisons, and the military, all serving as control centers for power, standardizing and regulating human behavior to maintain order and sustain the existing power structure. With the Industrial Revolution and the fossil fuel era, the amount of energy available to societies exploded, expanding the scale and complexity of social structures. This has led to unprecedented social complexity, with a highly detailed division of labor and global trade networks functioning as refined channels for managing energy and entropy. Modern systems, described by Foucault as *biopower*, monitor and regulate people's lives through public health policies, education, and economic control, emphasizing the management of life to optimize social energy flow. Michel Foucault's thought invites critical reflection on power dynamics within modern societies that, although they have achieved unprecedented success in increasing and managing energy, often perpetuate and exacerbate deep-seated inequalities. The critique extends to how economic and social structures modulate our living, prompting us to consider how we can redefine and reform our systems to encourage more just and equitable access to and control over energy resources, promoting a more ethical distribution of energy and a more balanced management of social entropy.

The ecological modernization theory suggests technological innovation may be the key to achieving more sustainable energy production and consumption patterns. However, the transition to this longed-for sustainability faces significant challenges. Current energy consumption patterns threaten the planet's ecology. At the same time, growing income inequality and the cultural upheaval triggered by staggering technological change challenge social cohesion.

The phenomenon of the *Nimby syndrome* ("not in my backyard") embodies resistance to change and the difficulty of dealing with the negative consequences of energy and technology, reflecting a tendency to reject the negative side effects of innovations, even when their benefits are widely accepted. Resistance to change is one aspect of social entropy that must be understood and addressed to move toward a more sustainable future. The history of the evolution of our societies is one of continuous adaptation to increasing energy flows and rising entropy. As we look to the future, it becomes imperative to use thermodynamic principles to shape a sustainable social path that respects the limits of our environment and promotes equity and social justice. In short, we are faced with the need for a new model of civilization that balances progress and conservation, growth and responsibility, innovation and tradition, power and participation. A model that reproduces a sense of belonging to the Earth on a planetary scale with a collective commitment to sustainability, equity, and resilience.

7.6 Rise and fall of civilizations

The concept of *catastrophic complexity*, derived from Joseph Tainter's theories, explains how societies can collapse under the weight of ever-increasing complexity, especially when insufficient energy is available compared to the energy required to maintain that complexity. Civilizations, like all things, are subject to the laws of thermodynamics. They rise and fall as they move through the interplay of energy and entropy, characterizing human history with a fascinating cycle of growth and decline.

Civilizations come into being when they find efficient ways to harness and use energy, whether in the form of fertile land, abundant natural resources, or innovative technology. Surplus energy enables the development of complex social structures, arts, sciences, and other hallmarks of civilization. The legendary city of Machu Picchu, built atop mountains, is an example of how geography and energy engineering influence civilization, as its inhabitants harnessed hydropower and terraced agriculture to sustain their population. Yet, societies die when the balance between resource consumption and resource renewability fails. Collapse occurs when entropy, or systemic disorder and energy dissipation, exceeds a civilization's ability to regenerate and maintain needed resources effectively. History teaches us that the fate of advanced civilizations such as the Maya or Easter Island was sealed when unsustainable practices exceeded the resilience of their ecosystems, leading to famine, conflict, and, eventually, decline. In a relentless cycle of rise and fall, civilizations are constantly challenged to innovate and adapt in ways that preserve harmony between their needs and the limits of the natural world.

In his book *Collapse*, Jared Diamond examines how poor environmental decisions can contribute to civilizations' downfall, showing that sustainable resource management is crucial to long-term maintenance. However, a civilization's growth increases its complexity and, consequently, its entropy. Maintaining order in the face of entropy requires energy to enforce laws, manage resources, coordinate activities, and so on. Therefore, the rise of a civilization can be seen as a process of continually increasing the use of energy to manage increasing entropy.

The phenomenon can be further explained by *Complex Adaptive Systems Theory*, which states that civilizations behave like living systems, self-organizing and adapting as long as resources allow. Eventually, many civilizations reach a point where the energy required to manage their complexity exceeds their ability to harness energy. All civilizations face an energy deficit, whether due to resource depletion, environmental changes, social unrest, or external threats. Unable to maintain their complexity, they decline and eventually collapse, in a manifestation of the inexorable rise of entropy.

In a broader perspective, philosopher Spengler, in *The Decline of the West*, sees the cycle of rise and fall as a *cultural fate*, almost like a season of civilization that follows an inevitable course, regardless of individual or collective actions. His historical fatalism refers to a kind of biology of civilizations, where each goes through the stages of birth, growth, decline, and death, similar to living organisms.

Let's consider the Roman Empire, whose rise was fueled by conquests that increased its wealth (a form of energy). As the empire grew, it faced increasing entropy from internal conflicts, economic crises, and pressure from external invaders. Despite their best efforts, the Romans could not muster enough energy to manage the growing entropy, leading to the decline and fall of the empire. The modern globalization process, with its interconnectivity, could be compared to the Roman Empire in terms of complexity and energy challenges, as Thomas Homer-Dixon notes in *The Upside of Down*. Such a perspective does not paint a fatalistic picture. Still, it serves as a reminder of the balance that must be struck between energy use and entropy management in maintaining any complex system, including civilizations. It calls for sustainable energy practices, thoughtful resource management, and careful societal planning as we forge our civilization's path into the future.

Nassim Taleb's concept of *antifragility* may offer a glimmer for future societies, in which it is not enough to resist disruptions but to evolve and improve according to them, turning crises into opportunities for renewal and growth. Taleb argues that systems that can be defined as antifragile benefit and become more potent when exposed to volatility, randomness, and disorder. This implies the creation of societies capable of adapting and thriving in response to unforeseen stress and change, optimizing resources and structures for inevitable uncertainty. In a world where the only constant is change, antifragility becomes a critical guiding principle for sustainable progress, pushing civilizations to survive and develop strategically in response to emerging challenges.

7.7 War and Peace

War and peace are two states of social existence that reveal different thermodynamic signatures. In its most basic sense, war is a high-energy, high-entropy event. It requires an enormous expenditure of energy, not only in terms of weapons and logistics but also in terms of human costs, infrastructure, and disruption of social norms. The chaos and unpredictability of war are manifestations of increased entropy.

Studying historical wars, such as the two world wars, we observe that the cycle of civilization building and destruction follows a thermodynamic rhythm, where post-conflict reconstruction often leads to increased order and innovation, as demonstrated by the Marshall Plan in Europe.

In the heat of conflict, war can seem like a challenge to the second law of thermodynamics. Energy appears concentrated, focused on destruction rather than dispersion, as in a temporary aberration. Over time, the energy expended during war eventually dissipates, manifesting itself in physical destruction, loss of life, social instability, and economic decline. The apparent order of military organization gives way to the disorder of violence, loss, and instability.

An aspect well exemplified by Schumpeter's concept of *creative destructiveness*, which, although generally applied to economics, can also be extended to cycles of war and peace.

Peace, in contrast, is a state of lower energy and entropy. It allows power to be conserved by focusing on constructive endeavors such as infrastructure development, scientific progress, and societal welfare. Peace's stability and predictability also represent a state of lower entropy, promoting the growth and development of social complexity. Peace can also be analyzed through *game theory*, where cooperation between parties can lead to stable order and mutual benefit, reducing the entropy of the overall social system.

In this context, peace can be seen as a social mechanism for managing and reducing entropy, ensuring stability, and promoting the efficient use of energy. It facilitates energy flow toward maintaining and improving social structures, nurturing human potential, and advancing knowledge and innovation.

Nobel Peace Prize winner Muhammad Yunus has shown that innovative solutions such as microcredit can act as catalysts for positive social energy, promoting peace and prosperity through the economic empowerment of the most disadvantaged. It is a kind of initiative that highlights the transformative power of the social economy, where small amounts of capital can unleash human potential and create waves of change that extend far beyond the individual. With microcredit, people who would otherwise remain on the margins of society can start businesses, educate their children, and improve their quality of life, thus starting a virtuous cycle of development and growth. It is a striking example of how economic energy can be distributed more equitably, sparking meaningful and sustainable change.

However, it is essential to remember that neither state is inherently sustainable without proper energy management. A peaceful society that fails to manage its resources efficiently may face internal conflict or external threats. Likewise, a society at war risks expending all its energy in conflict, leading to eventual collapse. Finally, the principle of *sustainable peace*, which integrates environmental, economic, and social management, becomes critical to mitigating social entropy and promoting a lasting order.

Understanding the thermodynamics of war and peace provides valuable insights into how societies can strive for stability and growth. It emphasizes the importance of sustainable energy practices, social equity, and diplomacy in managing entropy.

7.8 Globalization

Globalization, in its essence, is a manifestation of the second law of thermodynamics. It involves the flow of goods, services, information, and people across borders, reflecting energy flow and entropy dispersion in a physical system. Global expansion demonstrates the principle of maximum entropy, where systems naturally tend toward maximum entropy or disorder. Globalization is the social and economic process that follows this principle, spreading and culturally mixing populations similarly to how heat spreads in a closed system. From a thermodynamic perspective, globalization represents a state of increased entropy in human society. Boundaries dissolve, information is shared, and goods and services flow freely, reflecting the dispersion of particles in a closed system in search of equilibrium. The more societies intermingle, the more consequently the complexity and diversity of the world will increase, just as the increase of entropy in a system occurs.

Economist Thomas L. Friedman, in his book *The World is Flat*, explains how globalization flattens hierarchies and creates a more level playing field, increasing competition and interdependence. Globalization, through widespread access to technology and information, is shrinking the distance between individuals and markets. This results in a leveling of opportunities and information that used to be the preserve of the few, pushing toward cultural and economic homogenization. In this new landscape, ideas spread with astonishing speed, generating unprecedented innovation and interaction across cultures. At the same time, such openness has exacerbated the visibility of inequalities and raised new questions about sovereignty and cultural identity. Moreover, globalization can be seen as a heat engine on a global scale. Economic disparities between developed and developing countries act as a temperature difference, driving the flow of resources and energy. Wealth and technology flow from hot (rich) regions to cold (poor) regions in an attempt to achieve balance. Just as in a heat engine, the flow will always be imperfect, and inefficiencies will always exist, often leading to wealth accumulation and economic imbalances. This process can be compared to the Carnot cycle in the economic context, where efficiency is related to the ability to convert energy (resources) into helpful work (economic development and social progress). The mechanism of globalization, reflecting the *Carnot cycle*, occurs through a process of exchange not without energy losses, manifested through socioeconomic inequalities. More developed nations, similar to a high-temperature heat reservoir, transfer technology and investment to poorer countries, which can be likened to a low-temperature reservoir. Although aiming for greater global economic homogeneity, energy transfer often does not result in a balanced benefit because of structural, cultural, and political factors that influence the absorption and use of these resources. The result is a complex web of trade that, while on the one hand, fuels growth and innovation, perpetuates and sometimes exacerbates imbalances. The resulting economic interdependence can generate a system in which inefficiencies, rather than being dissipated, become concentrated, giving rise to cycles of poverty and dependence that run counter to the ideal of shared progress promised by globalization.

Information technology acts as an entropy accelerator. By enabling real-time, worldwide communication and data sharing, it breaks down barriers and accelerates the global dispersion of information, ideas, and innovations. This accelerated flow creates greater complexity but also greater potential for innovation and progress.

Moravec's paradox highlights a fundamental shift in work and society. While complex cognitive tasks are digitized easily, challenging the value of the human intellect in the labor market, trades requiring manual and sensory skills remain challenging to automate. This will lead to a new form of labor and social entropy, where the value and allocation of human labor will be redefined. Automation and artificial intelligence are not just tools of efficiency; they are forces that reshape power dynamics, wealth distribution, and the very structure of communities. Globalization accelerates this process as labor and skills move with unprecedented speed, often leaving communities shaken by technological unemployment and growing economic inequality behind.

Nevertheless, globalization also poses challenges. Increased complexity can lead to confusion, conflict, and social instability. In addition, the current globalization model relies heavily on energy-intensive industries and transportation, exacerbating our environmental challenges. Globalization has woven a web of interdependencies that, while promoting economic and cultural integration, carries the seeds of possible friction with it. Globalized industries and transportation, pillars of contemporary development, burden our environment with their demands for fossil energy, increasing greenhouse gas emissions and exacerbating the problem of global warming. A dynamic that raises questions about the long-term sustainability of our current economic model and the responsibilities each nation has to the planet.

The notion of *degrowth*, proposed by Serge Latouche, can offer a thermodynamic solution to the problem, advocating an economy that aims to reduce its energy and material *throughput*. Latouche proposes a radical rethinking of economics, suggesting that *less is more*. Degrowth does not mean regression but a redirection of energies toward more environmentally friendly and socially just practices, implying a detachment from the obsession with GDP and embracing values that promote quality of life and environmental protection.

In the future, the thermodynamics of globalization implies the need for balanced growth, equitable resource distribution, and sustainable energy use. It involves fine-tuning the thermal engine of the global economy to minimize inefficiencies, reduce entropy spikes, and distribute energy more evenly, creating a more stable, sustainable, and integrated global society. That is, global policies must adopt strategies that incentivize technological innovation for energy efficiency and encourage a cultural shift toward responsible consumption patterns. The vision of an optimized heat engine in the global economic context invokes a transformation that views energy not as a commodity to be exploited without limits but as an asset to be managed wisely.

The ultimate goal will be to reach a *triple point of sustainability*, where economic, social, and environmental balance coexist harmoniously, similar to the triple point of water in physics, where the solid, liquid, and gas phases coexist. Achieving this triple point requires fresh thinking and an innovative approach to economic, social, and political activities so that they are in tune with the natural rhythms of our planet. The vision is of a world in which renewable energy, social equity, and economic prosperity come together in a dynamic and sustainable balance that is in harmony with the inherent logic of thermodynamics and respect for life on Earth.

7.9 The thermodynamics of the Anthropocene.

The *Anthropocene*, the current geological epoch in which human activity is the dominant influence on climate and the environment, presents a unique thermodynamic perspective. It is an epoch of significant energy fluxes and dramatic increases in entropy triggered by human activity. It highlights humanity's unprecedented ability to influence the Earth system, but with it also comes an urgent responsibility to control the unintended consequences of that impact. As we exploit Earth's resources to feed our growing societies, we must face the reality of a planet with finite limits and the need to balance our aspirations with the sustainability of our environment.

Since the advent of the Industrial Revolution, humankind has harnessed and used large amounts of energy, mainly in the form of fossil fuels. Such energy use has generated significant technological advances and increased the complexity of human societies; nevertheless, it has also led to a substantial increase in entropy. Pollution, climate change, and biodiversity loss are physical manifestations of this increase.

According to physicist and historian of science Spencer Weart, the Anthropocene presents an unprecedented challenge in the history of science: to control the entropy that we ourselves have unleashed. Scientific awareness and responsibility must guide our efforts in reducing humanity's entropic footprint. Weart's research calls for unprecedented interdisciplinarity, bringing together scientists from different disciplines to address the effects of the entropy we ourselves have generated, seeking innovative and sustainable solutions.

Human-induced climate change is particularly illustrative of the thermodynamic principles at play in the Anthropocene. Burning fossil fuels releases energy, but it also produces carbon dioxide, which traps heat in Earth's atmosphere. The added heat disrupts

Earth's climate system, creating more energetic weather patterns and increasing the system's overall entropy. The effects of global warming are a clear demonstration of how human activities can alter the thermodynamic principles that govern Earth's climate. Thus, a new equilibrium is needed, where energy consumption aligns with Earth's ability to absorb and recycle energy in ways that do not destabilize delicate climatic and ecological systems.

This situation is similar to what physicist Jeremy England describes as *guided dissipation*, where biological and ecological systems adapt to dissipate energy more efficiently in response to external forces. England reminds us that solutions to environmental problems must be as adaptive and dynamic as the systems we try to preserve. By understanding and applying principles such as guided dissipation, strategies can be developed that address the symptoms and root causes of environmental entropy.

The loss of biodiversity – another critical problem of the Anthropocene – can also be analyzed through the lens of thermodynamics. Each extinct species is a lost piece in the great mosaic of life, the absence of which can irreversibly alter the energy flow and entropic dynamics of entire ecosystems. This phenomenon mirrors Elizabeth Kolbert's *sixth mass extinction theory*, which illustrates how biodiversity loss can lead to a thermodynamic point of no return for Earth's ecosystems. Kolbert warns us that each extinction brings us closer to an ecological breaking point, accelerating our entry into an era of biological and thermodynamic uncertainty.

Yet, the Anthropocene should not only be understood as a story of increasing entropy. It is also a testament to human ingenuity and our ability to understand and manipulate our planet's energy flows and entropy. The rise of renewable energy technologies, efforts at reforestation and conservation, and the development of strategies to combat climate change are all examples of humanity's attempts to counterbalance the increasing entropy of the Anthropocene. Humanity's innovative efforts in this era are a testament to our innate desire to create and maintain order in a world subject to constant change. Initiatives toward sustainability, from clean energy production to biodiversity conservation, show our growing awareness of the need to live in harmony with the thermodynamic laws of the planet.

The concept of a *circular economy* is fundamental to the attempt to close the cycle of matter and energy flows to minimize waste and entropy. Circularity in the economy is the recognition that the linear model of production and consumption is unsustainable. By integrating the concept of a circular economy into our policies and practices, we can reduce our entropic (and anthropogenic) impact while conserving resources for future generations.

The future of the Anthropocene will be determined by the ability to manage energy flows and entropy increases. Success will depend on innovation, cooperation, and a deep understanding of our world's thermodynamic principles. Our ingenious challenge in the face of entropy will shape our future in the Anthropocene. Transformation to a sustainable society requires a clear vision and concerted action. The transition to sustainability can be likened to reaching a *critical point* in physics, a phase change necessary to enter a new era of thermodynamic equilibrium between humanity and nature. The crucial point is a metaphor and a tangible goal, requiring a collective shift in our consumption, production, and thinking patterns. Only then can we hope to enter a new era of dynamic equilibrium in the Anthropocene.

7.10 The future of societies

Thermodynamics provides a framework for anticipating potential paths in predicting the future of societies. It suggests that a society's success depends on its ability to balance the interplay of energy acquisition, utilization, and dissipation against the backdrop of entropy and complexity. Looking ahead, societies will continue to face the challenge of managing increasing complexity. As the population increases and technological advances accelerate, the entropy of our social systems will also increase.

Managing entropy efficiently while continuing to meet the energy demands of growing and evolving societies will be one of our most significant challenges. Our resilience and adaptability will be tested in an age defined by the rapidity of change and the intensity of global connections. Renewable energies, such as the sun and wind, flowing abundantly and inexhaustibly above us, hold the key to unlocking a sustainable future. Such energy flows, harmonized with technological innovation, offer a promise of balance in which energy production is no longer a threat but an opportunity to reduce environmental entropy and sustain growth. Renewable energy sources will play a vital role in this endeavor. As fossil fuels become scarcer and climate change intensifies, the shift to sustainable and renewable energy will be necessary not only for environmental reasons but also for the survival and growth of our societies. The ability to harness and store energy from renewable sources efficiently will define the future trajectory of civilizations.

The transition to renewable energy is an ecological imperative and a social mandate. Equity in energy access and equitable distribution of resources will become pillars of a prosperous society, reducing inequalities that can destabilize the very foundation of our social interactions and increase human entropy.

At the same time, managing the entropy of our social systems requires a particular focus on equality and justice. Disparities in wealth and resources can lead to social instability, a form of social entropy. The more complex societies become, the more critical it will be to address inequalities in order to maintain order and cohesion.

Technological advances will also shape society's future. Innovations in fields such as artificial intelligence, robotics, and biotechnology could significantly alter the way we live, work, and interact. They could also increase society's complexity, necessitating a careful balance between technological progress, energy use, and entropy management.

The future of societies depends on our ability to sustainably manage energy and complexity in a multifaceted approach that combines technological innovation, renewable energy, social justice, and efficient *governance*. Ultimately, there is a synergy between energy engineering, social architecture, and environmental ethics.

From the Big Bang to Consciousness

From the Big Bang to Consciousness

Notes and bibliographical insights in Chapter 7

1. Ilya Prigogine, *Order Out of Chaos* (1984). Prigogine explores the concept of dissipative structures and systems far from equilibrium, offering a unique perspective on how societies can maintain and develop order in an entropic environment.

2. Leslie White, *The Science of Culture* (1949). White introduced the energy value theory, arguing that social complexity is directly proportional to the amount of energy per capita controlled by a society.

3. Émile Durkheim, *The Division of Labor in Society* (1893). In this work, Durkheim explores how anomie, or the lack of norms, emerges in rapidly changing societies and can be linked to growing social entropy.

4. Niklas Luhmann, *Social Systems* (1995). Luhmann applies complex systems theories to sociology, proposing that societies function as autopoietic systems that maintain their structure despite destabilizing external influences.

5. Jared Diamond, *Guns, Germs, and Steel* (1997). Diamond examines how societies develop and dominate based on available energy and technological resources, directly linking energy and social structure.

7. Joseph Tainter, *The Collapse of Complex* Societies (1988). Tainter discusses how complex societies can collapse under the weight of their own complexity when the energy required to maintain that complexity exceeds the energy available.

8. Manuel De Landa, *A Thousand Years of Nonlinear History* (1997). De Landa provides an interdisciplinary view of human structures using concepts from thermodynamics and complex systems theories.

9. Gregory Bateson, *Steps to an Ecology of Mind* (1972). In this book, Bateson explores how systems of ideas, including societies, emerge through

interactive processes. These processes can be analyzed using thermodynamics and information theory principles.

10. Richard Dawkins, *The Selfish Gene* (1976). Dawkins introduces the concept of memes, replicating units of cultural information analogous to genes in biology, which can be used to explain the evolution of social structures.

Chapter 8

TECHNOLOGY:
HUMANITY'S THERMODYNAMIC TOOL

From the Big Bang to Consciousness

From the Big Bang to Consciousness

Between intellect and intuition
there is a bridge,
which is the experience

CARL JUNG

Psychologische Typen (1921)

From the Big Bang to Consciousness

8.1 The dawn of technology

Technique – understood in the broadest sense – is the method humanity has developed to interact with and manipulate the world around us. Its beginnings can be traced back to the earliest human ancestors, who used simple tools such as sharp stones and pointed sticks to facilitate survival. Such tools were nothing more than a rudimentary way of manipulating energy flows and entropy in the environment to achieve desired results, from hunting and gathering to building shelters.

Heidegger, in contemplating technique, looked beyond physical tools. He saw technique as a *manifestation of being*, a way of opening up the world to make it *available, manipulable, and exploitable*. An unveiling, however, that also entails a domination of the natural environment, a concern even more pertinent today. Most importantly, such a vision makes technology the *subject* and humans the *object of* it. In this reversal of roles, technique, from a mere extension of human capabilities, is transformed into an autonomous force that shapes humanity in its own image. The human, therefore, finds himself in the condition of being shaped and defined by the tools he has created, falling into the trap of a dependence that limits his freedom and capacity for self-determination. The challenge is, therefore, twofold: to recognize and respect the autonomy of technology without succumbing to it and to rediscover a balance in which a human guides it toward ends that enhance and preserve the essence of nature and of the human being himself. Only in this way can we hope that technique will remain a means to improve the human condition and not an end that moves us further away from our authentic being.

Myth has addressed this conflict and the ensuing perpetual quest for balance through the figure of Prometheus, who steals fire from the gods to give it to humanity, embodying the boldness and revolution that controlled access to energy has meant for our species. From ancient myth emerges the theme of energy as a catalyst for human progress, an early symbol of human ascendancy through technology.

Fire, the first *human-controlled bioenergy*, not only marked the beginning of a new way of life but also laid the foundation for the first significant step toward civilization and industry. With the discovery of fire, we gained the ability to transform materials and energies, ushering in the era of technological innovation. Despite their immediate benefits, the first technological steps had increasing costs of unanticipated entropy. While a fire flame provided warmth and protection, it also consumed resources and, over time, contributed to deforestation and altered ecosystems. The path of human innovation has always had to deal with the consequences of changing the surrounding environment. Thermodynamically, the evolution of prehistoric tools improved efficiency in converting energy into useful work, such as cutting meat with sharp stones. The increased efficiency reduced local entropy, allowing humans to achieve more with less energy. As civilization progressed, the energy efficiency of tools expanded human capabilities, overcoming the limits imposed by our biology and paving the way for human-defining forms of cultural and intellectual expression.

Today, the concept of *ecological footprint* provides a yardstick for assessing the environmental impact of modern technology. Whereas our ancestors' tools were measured in terms of their immediate efficiency, we now have to calculate the total cost of their use while also considering their long-term impact on our planet's entropy.

8.2 The role of energy in technological progress

The progress of technology is intrinsically linked to humanity's ability to harness and manipulate energy. This relationship becomes evident when we look at significant milestones in human history. For example, the harnessing of fire – the control of heat energy – for the protection, cooking, and processing of metals. Fire not only provided heat but also initiated a chain reaction of innovations, such as the production of ceramics and the processing of metal alloys, demonstrating the catalytic effect of energy on technological progress.

The *Agricultural Revolution* marked another significant change, focusing on efficiently harnessing the sun's energy. By domesticating plants and animals, early humans could secure a more reliable source of food by translating the sun's energy into food energy through agriculture rather than relying on the randomness of hunting and gathering. All this allowed for surplus food production, which led to population growth and the creation of more complex social structures. Regarding thermodynamics, agriculture turned solar energy into biomass much more directly and efficiently than harvesting or hunting, optimizing the use of available solar energy. The Agricultural Revolution also introduced new social and environmental dynamics. The centrality of agriculture in human societies not only transformed the landscape but also influenced the structure of communities. The accumulation of resources set the stage for early economic inequality and social power forms. In this context, agriculture increased energy efficiency and introduced new forms of social entropy, such as struggles over resources and surplus management.

The *Industrial Revolution* brought a new form of energy use with the advent of the steam engine and other machinery powered by fossil fuels. Coal, oil, and gas became the primary energy sources that fueled unprecedented technological growth, manufacturing, transportation, and more. The ability to extract and use solar energy stored within fossil fuels has created a rapid surge in our ability to work and reduce local entropy, leading to massive social changes. The era of fossil fuel combustion can be defined as a period of high energy use and increasing global entropy, a dynamic underscored by the *tragedy of the commons* described by Garrett Hardin. During the Industrial Revolution, the massive use of fossil fuels led to an unprecedented increase in environmental pollution and climate change. As industrialized societies prospered, the consequences of unsustainable energy use began to emerge, highlighting the importance of balancing energy exploitation with environmental sustainability, a concept that continues to be critical in the modern era.

In the *Information Age*, electricity and, more recently, renewable energy sources have become critical. These forms of energy power our computers, data centers, and Internet infrastructure, enabling the digital technologies that define our modern world. In addition, the rise of renewable energy technology, such as solar panels and wind turbines, reveals our growing ability to harness energy directly from the environment more sustainably. The Information

Age has brought with it new challenges and opportunities. While digitization has made information and communication more accessible, it has also intensified global energy consumption. The growing dependence on digital technology raises questions about the long-term sustainability of such growth and the environmental impact of data centers and digital infrastructure. Shifting to renewable energy is essential to reduce environmental impact and maintain sustainability.

The challenge will be to balance technological innovation with energy and environmental sustainability in the future. Technological advancement must be accompanied by increased ecological awareness and strategies to reduce environmental entropy. Future innovations must be evaluated by their technological impact and ability to contribute to a sustainable and energy-efficient world. This balance is essential to addressing the *great acceleration* of the Anthropocene and ensuring that the benefits of technological progress are not obscured by their environmental consequences, as illustrated by sustainability and environmental science research.

8.3 The thermodynamics of innovation

Innovation, the process of creating new solutions to problems, can be seen as a mechanism for managing and manipulating entropy. When facing challenges, innovations are needed to reduce disorder in local systems, push entropy outward, and create pockets of relative order. This is what all technologies aim to achieve.

Philosopher Karl Popper proposed the concept of *evolutionary epistemology*, suggesting that knowledge grows through conjecture and refutation, similar to the scientific method. It is an approach that also applies to technological innovation, where each error becomes a step toward a more orderly and lower entropy solution. Innovation should, therefore, be experienced as a dynamic process, a continuous dialogue between theory and practice, where each new discovery or invention leads to new questions and challenges. A continuous iteration of ideas and solutions leads to a perpetual cycle of improvement and optimization, progressively reducing entropy in the systems and technologies we build.

Consider the innovation of the wheel, which significantly reduced the entropy associated with moving objects over a distance. Before this invention, the effort and energy required to transport goods was immense, resulting in a high entropy cost. The wheel enabled a more efficient energy transfer, reducing the overall entropy generated during transport. Not only did it simplify physical transportation, but it also paved the way for a whole range of mechanical and industrial applications. It ushered in an era when machines could perform tasks that previously required considerable human or animal effort, further optimizing energy use and reducing entropy in production and handling processes. The wheel not only transformed transportation but also initiated the development of complex machines. The energy expended to turn the axle of a wheel was the precursor to more complex mechanisms, such as gears and transmissions, which form the basis of modern mechanical engineering.

The very process of innovation can be seen as a form of entropy management. It involves going through a series of ideas, many of which lead to dead ends, while only a few offer viable solutions. One concept that aligns with the notion of *adjacent possible* – a term coined by theoretical biologist Stuart Kauffman. It refers to the set of all first-order variations on the current state of a system – the possibilities that lie at the edge of our current understanding or capacity. The concept suggests that innovation advances through small steps rather than giant leaps. Each new invention or discovery opens up new possibilities, expanding our horizon of what is achievable. Innovation is thus an evolutionary process, with each new idea building on the foundation of previous ones. It allows us to combine different perspectives and skills to find innovative solutions that address the multidimensional challenges of today's globalized society.

The adjacent possible also expands through the interaction and convergence of different disciplines, a phenomenon known as *cross-innovation*, where solutions in one field can suddenly solve problems in another, decreasing entropy in unexpected ways. Constant and iterative exploration is the key to understanding and harnessing the latent potential of our environments and technologies. It guides us toward greater knowledge and skill in managing and manipulating energy and entropy, leading to the realization of increasingly efficient and sustainable solutions. Exploring the adjacent possible, we enter the realm of higher entropy (disorder and uncertainty) in search of new, lower entropy (order and utility) solutions. It is a process of trial and error, full of failures and iterations, reflecting evolution's approach to creating increasingly complex organisms.

The process is accelerated by digitization and computational capacity, as described by complexity theorist Sam Arbesman in his book *The Half-life of Facts*, where computer technology has exponentially increased our ability to explore new solutions and knowledge. The digital age has transformed the way innovation happens, leading to an acceleration in the cycle of experimentation and learning. The ability to rapidly analyze large amounts of data and simulate complex scenarios allows ideas to be tested in ways that were not previously possible. The accelerated innovation process allows for faster exploration of the adjacent possible and optimized solutions.

The thermodynamics of innovation also underscores why technology progresses not linearly but exponentially. Each successful innovation opens up a new set of possibilities – the adjacent possible expands, leading to more potential paths to decrease entropy. Thus, technology builds on itself, with each advance providing the foundation for the next. The same pattern has been maintained from the development of simple tools millions of years ago to the most cutting-edge AI (*Artificial Intelligence*) systems. In this view, innovation is an evolutionary process where each new technology changes the landscape of possibilities, allowing new ideas and solutions to emerge. The expansion of the adjacent possible is the engine of exponential progress, driving humanity toward a future of ever more advanced technologies and ever more efficient systems.

As sociologist Helga Nowotny cautions, however, innovation must be balanced with ethical considerations and reflections on its long-term sustainability to ensure that reducing local entropy does not lead to increasing global entropy. Innovation must occur responsibly; that is, it must be accompanied by a deep understanding of its environmental, social, and economic implications. The challenge lies in creating technologies that not only solve the immediate problems of the present but also do so in a sustainable and beneficial way for future generations. The *holistic view of innovation* emphasizes the need for a balanced and informed approach that considers the long-term consequences and well-being of the entire planetary ecosystem.

8.4 Technology and energy dissipation

The evolution of technology has been about increasing our ability to work or create order and managing the energy we dissipate in the process. The relationship becomes apparent when considering energy efficiency principles in designing and operating tools and machinery.

For example, the principle of energy efficiency extends to the law of conservation of energy, indicating that while energy cannot be created or destroyed, it can be transformed from one form to another with various efficiencies, a fundamental concept expressed by the first principle of thermodynamics. Perfectly synchronized with the relentless pace of progress, each technological innovation looms as a mirage of efficiency in the growing desert of entropy. With energy flowing through our world like blood through veins, every device and every machine we have built is a tribute to humanity's daring attempt to tame and channel its power.

The steam engine, a hallmark of the Industrial Revolution, is a prime example. Although early designs were relatively inefficient, as much of the energy from the burning coal was lost as heat, subsequent improvements dramatically increased its efficiency. The development of condensing engines and later high-pressure and compound engines made it possible to convert more energy into useful work. With the evolution of steam engines to modern combustion engines and turbines, the second principle of thermodynamics has been applied, which states that the natural direction of energy is to move from a state of order to a state of disorder, or entropy.

Humans have demonstrated a consistent zeal to dominate and use energy to their advantage, giving rise to an era in which productivity and progress are measured by the ability to reduce entropy in the places where we live and work. In contrast, far from our eyes, progress's actual thermodynamic cost continues accumulating in atmospheres and oceans. The race toward ever-increasing efficiency has profound consequences, driving scientists and engineers to seek innovative solutions in the extreme boundaries of physics and chemistry. Managing thermal entropy is no longer an isolated engineering challenge; it has become a living metaphor for the balance that must be struck between our desire for progress and the limitations imposed by the immutable laws of nature.

Energy dissipation has been a central concern, even in electrical and electronic devices. For example, a significant part of computer engineering concerns the management of heat generated by processors. The smaller and more powerful components become, the more challenging it becomes to dissipate excess heat. This demands a constant search for new materials with better heat dissipation properties (such as nanomaterials and advanced cooling fluids) and spurs innovation in cooling systems and thermal design.

In addition, the rise of renewable energy technologies, such as solar panels and wind turbines, represents an ongoing effort to capture and use energy that would otherwise be dissipated and lost. By harnessing the energy of the sun or the kinetic energy of the wind, the technologies convert some of the natural energy flow into electricity, a more valuable and manageable form of energy for our technological infrastructure.

The *urban heat island* concept further illustrates how technologies and urbanization have transformed solar energy and thermal entropy globally, affecting local and global climates.

In each case, the goal has been to minimize the energy lost to the surroundings (i.e., energy dissipated) and maximize the energy used for useful work. This balance is a fundamental aspect of thermodynamics and a critical factor in the evolution of technology. As we strive for ever-increasing efficiency, we must also be aware of the broader thermodynamic implications, including the increase in global entropy associated with our energy consumption.

The future challenge for technology will be integrating concepts such as life cycle analysis to fully assess a product's environmental impact from cradle to grave, ensuring that the pursuit of efficiency does not sacrifice sustainability.

8.5 The information age: a thermodynamic revolution

With the advent of the *Information Age*, the link between thermodynamics and technology has taken on an even deeper role. The concepts of entropy and information have become deeply intertwined, leading to an intellectual revolution that has reshaped our understanding of everything from computers to quantum physics. Once considered an intangible abstraction, information has turned out to have a tangible thermodynamic weight. Every bit of data, every logical transition in a circuit, has an equivalent in terms of energy and, by extension, entropy. An unexpected link has opened new frontiers in computation, suggesting how every computation has an actual physical impact, limiting our processing capabilities and the speed at which we can communicate.

Theoretical physicist Leonard Susskind explored such connections in the *battle for the black hole*, showing how information is not lost in black holes but rather intricately linked to physics' fundamental laws. This principle has deep resonances in computer science and thermodynamics. This revelation has implications beyond astrophysics, affecting how we conceive data storage. Information loss is not just a digital event but reflects a physical transformation, offering a revealing perspective on the inherent cost of our information technologies.

In the realm of technology, the transistor, an electronic device that can amplify or switch electronic signals and electrical energy, is a prime example. Invented in 1947, it has become the fundamental building block of modern digital circuits, including microprocessors. The ability to precisely control and direct the flow of electrons has minimized energy dissipation and made electronic computing much more efficient. The evolution of the transistor, from its primordial form to modern MOSFETs, embodies the relentless quest to reduce entropy in every energy exchange. The path of refinement reflects our struggle against thermodynamic irreversibility, in which we try to capture every useful fragment of energy before it escapes into the environment.

Advances in semiconductor materials have expanded the miniaturization and energy efficiency frontiers, creating simultaneously powerful and compact devices thanks to innovations such as silicon-on-insulator and nanostructures. The silicon domain has made it possible to build architectures at the atomic level, where every atom counts and every electron is guided along defined paths to minimize *wasted energy*. The drive for efficiency is not simply a quest for technological perfection but a necessary response to the limits imposed by thermodynamics.

The real thermodynamic revolution of the Information Age lies in our ability to create, store, and manipulate information. The very concept of information has strong links to entropy. In the mid-20th century, physicist and mathematician Claude Shannon demonstrated that information could be quantified in a manner analogous to thermodynamic entropy, creating the field of information theory.

Shannon's equivalence between bits and entropy is more than a mere analogy; it is a revelation of how our digital world is intertwined with the physical laws that govern the universe. Every time we delete a file or save a document, we affect our world's energy balance. The equivalence between information and entropy also manifests itself in modern data compression algorithms, where the goal is to reduce redundancy or entropy to represent information as efficiently as possible. Data compression is the art of extracting essence from excess and transforming chaos into coherence. In each compressed byte, there is a victory against disorder, a small triumph of order in the ocean of entropy.

Shannon's theory has profound implications for our understanding of information technology. In essence, when we store or transmit information, we create order – decreasing entropy at the local level. A hard drive storing a terabyte of data is a highly ordered system compared to one full of random bits. Yet, the ordering process is not free. It requires energy, both to write the data and to maintain it. And energy, in turn, contributes to a general increase in entropy elsewhere. As data centers multiply and the cloud expands, the energy used to keep our digital world alive becomes a growing concern. Every gigabyte of data, every search query, and every streaming video is a drop in the bucket of energy consumed and entropy produced.

These principles imply that our increasing dependence on digital technologies requires us to consider the energy sustainability of IT infrastructures, as illustrated in research on *green IT* and sustainable computing.

As computing advances into quantum computing and artificial intelligence, a sustainable paradigm becomes imperative. We cannot afford to pursue innovation at the expense of our planet, and every step forward must consider eco-thermodynamics.

In addition, information processing -the operations performed by a computer-also has a thermodynamic cost. The energy efficiency of these processes and their contribution to total entropy are areas of ongoing research, particularly as we approach the physical limits of *Moore's Law*. Every click, every swipe, and every moment spent online is not energy-neutral. With Moore's Law approaching its theoretical limits, we must face the reality of a costly calculation in economic and thermodynamic terms.

Emerging technologies, such as *quantum computing*, promise to rewrite information processing rules, potentially reducing computational entropy and increasing energy efficiency to unimaginable levels. If quantum computing fulfills its promises, we could witness a new era of computing and a new chapter in humanity's thermodynamic history, one in which our energy footprint is reduced. At the same time, our processing capabilities expand in ways we can now only imagine.

The *Information Age*, then, is a technological revolution and a thermodynamic one. It presents new challenges and opportunities for understanding the profound relationship between energy, entropy, and information. As we advance toward an increasingly digital future, thermodynamic considerations will continue to play a central role in guiding the evolution of technology. As the network expands, our responsibility grows with it. The ethical integration of artificial intelligence, privacy, and data security deliberations must all be considered in light of the energy and entropy that accompany digital advances. Our aspiration for efficiency must always uphold the values of accountability and transparency.

8.6 The thermodynamics of the Internet

As an extension of the *Information Age*, the Internet is a vast global network deeply rooted in thermodynamics. Every e-mail sent, video streamed, or web page uploaded involves countless thermodynamic processes. The flow of electrons in computer circuits, the generation of light pulses in fiber optic cables, and the cooling of massive data centers are fundamentally thermodynamic in nature. The interconnection between thermodynamics and technology reflects the fundamental principle that energy can neither be created nor destroyed but only transformed from one form to another. In this context, data transmission over the Internet transforms electricity into information, heat, and ultimately into new energy consumption.

The global Internet network can be compared to a sizeable metaphorical heat engine. According to the second law of thermodynamics, information is the *energy* flowing through the *system*, and energy dissipation is inevitable.

The Internet has profoundly changed how we communicate, work, learn, and entertain. But beneath these changes, how we consume and dissipate energy has also altered. Data centers, which house the servers that form the backbone of the Internet, consume large amounts of electricity. Energy is needed to power and cool the servers themselves, as the calculations they perform generate significant amounts of waste heat. Data center architecture reflects an ongoing struggle to balance data processing needs with minimizing energy consumption and environmental impact.

Data centers are metaphorically the *lungs* of the digital infrastructure, *breathing* in electricity and exhaling heat – a direct manifestation of the entropy generated by information systems. Efficient management of thermodynamic *respiration* is critical to reducing the Internet's carbon footprint and ensuring the long-term sustainability of our growing consumption of digital services.

According to some estimates, the information and communication technology sector includes the data centers, networks, and devices we use to access the Internet, use up to 20 percent of all electricity, and emit more than 5 percent of the world's carbon emissions by 2025. These estimates underscore the thermodynamic implications of our digital age and the urgent need for more sustainable practices.

The challenge is to optimize the architecture of data centers and Internet networks to reduce energy consumption. This task requires ongoing innovation, such as using renewable energy sources and recycling waste heat for other purposes. Initiatives such as using natural chilled water to cool servers or implementing more energy-efficient server architectures demonstrate the potential for significant improvements.

The Internet's thermodynamics also have implications for efficiency and sustainability. As the demand for digital services grows, so does the need for energy-efficient hardware, software, and network design. Research in these areas is ongoing, ranging from attempts to create energy-proportional computing systems, which scale energy use according to demand, to advances in cooling technology and server architecture. Such innovations improve energy efficiency and help reduce the entropy generated by the global Internet system.

In this context, the *Internet of Things* (IoT) concept takes on a new thermodynamic meaning. It connects billions of devices and sensors, all of which demand energy and contribute to system entropy. The IoT amplifies the challenge of efficiently managing energy, as each additional device introduces energy demands and increases the system's overall entropy.

The Internet is also a catalyst for a new kind of resource consumption: the consumption of information itself. Nearly infinite access to information and content can lead to digital overconsumption, characterized by endless *scrolling*, *binge-watching*, and an endless stream of notifications. Information overconsumption has its own thermodynamic cost, as each bit of data consumed requires energy to produce and transmit. Our growing dependence on streaming platforms, social media, and online services raises essential questions about the energy footprint of our data consumption.

Similar to the concept of *digital obesity*, information overload and continuous demand for bandwidth can be seen as forms of overconsumption that require critical reflection on their energy impact and social entropy. The metaphor highlights the need for greater awareness and responsibility in digital consumption, promoting practices that reduce the Internet's energy consumption and ecological footprint.

The Internet is a testament to our ability to create and manage large-scale, complex systems. However, it also highlights the fundamental thermodynamic constraints that must be addressed as we continue to expand and rely on digital infrastructure. Understanding and innovating in response to such constraints is essential to maintaining the Internet as a sustainable and accessible resource for future generations.

Recognizing and addressing these constraints is essential to ensuring that the Internet's growth is sustainable, not only in terms of physical infrastructure but also in terms of environmental and social impacts. The transition to more sustainable Internet use requires a holistic approach that considers not only technologies and infrastructure but also consumption patterns and cultural practices. The challenge is complex, but the opportunities for innovation and improvement are immense. They offer the chance to reshape the digital landscape in ways that respect thermodynamic principles and promote a more sustainable future.

8.7 The energy cost of technology

As we move deeper into the Information Age, the energy cost of our digital world continues to rise. Every search we do, every email we send, and every video we stream all add up to a significant amount of energy consumption. And with the growing reliance on technology for every aspect of our lives, from work to education, entertainment to communication, the energy demand is bound to increase, raising crucial questions about the environmental impact and long-term sustainability of our digital habits. Reflecting on the history of energy consumption, we can see an uncanny parallel with the evolution of the Industrial Revolution, where increased efficiency led to greater overall resource consumption, a phenomenon known as the rebound effect.

Koomey's constant observes that computer energy efficiency roughly doubles every 1.5 years, suggesting an improvement in energy management and underscoring the increasing overall energy demand for computer systems. This trend highlights how resource consumption can increase rather than decrease overall use. The challenge is, therefore, twofold: to improve efficiency while moderately reducing energy demand.

Consider, for example, the energy consumption of data centers, which are the Internet's heart. These buildings house thousands of servers that store and process the enormous amount of data we generate daily. Cooling the servers to prevent overheating constitutes a significant part of energy consumption. According to some estimates, data centers worldwide consume more electricity annually than in some countries.

The challenge is not only technical but also systemic. Sustainable data center design involves integrating renewable energy and innovative cooling strategies, such as using natural cold water or direct immersion in nonconductive liquids. Such approaches require a systemic view that considers thermodynamics, technology, and ethics, promoting a synergy between technological progress and environmental protection.

But it is not just about the Internet. The devices we use to access the digital world – smartphones, laptops, smart TVs, IoT devices – also have significant energy costs, both in terms of production and use. In particular, the extraction and refining of rare earth elements and other materials used in the devices mentioned above represent a substantial energy expense. The exact energy cost of e-waste disposal and recycling must be added to the total energy footprint of the devices. In this sense, the principle of extended producer responsibility (EPR) could provide a framework for addressing the life cycle of electronic devices by holding manufacturers accountable for the ultimate fate of their products.

Every discarded device carries a shadow of entropy that never completely disappears; the *sunk cost* of technology includes the energy expended during use and that associated with its entire life cycle. Such a concept is reflected in theories of sustainability and the circular economy, which emphasize the need to reduce, reuse, and recycle to minimize environmental impact. The production of electronic devices, in particular, involves the extraction of rare metals, which is often associated with severe environmental and humanitarian problems. Recent studies indicate that the life cycle of a single smartphone can generate up to 60 kg of CO_2 emissions, equivalent to about 280 km traveled by an average car.

Even seemingly minor aspects of our digital lives can have surprisingly high energy costs. For example, video games can be a highly energy-consuming activity as we move toward using increasingly high-resolution screens and more powerful graphics processors. Similarly, the energy consumption of cryptocurrency mining operations has become a topic of growing concern. These phenomena reflect the growing demand for data and computationally intensive computing, with data centers around the world consuming about 2 percent of global energy, a figure that is expected to increase in the coming years.

The energy footprint of a single Bitcoin transaction is comparable to that of thousands of credit card transactions, illustrating how the new financial technology can have disproportionate energy costs compared to traditional systems. Therefore, research on blockchain and cryptocurrencies is moving toward less energy-intensive algorithms, such as *proof-of-stake*, that promise to reduce this impact significantly.

In all these cases, the energy cost of the technology represents a thermodynamic trade-off. On the one hand, the technologies offer enormous benefits in terms of productivity, convenience, and entertainment. On the other, the energy they consume and the heat they dissipate contribute to the overall increase in entropy in the universe, not to mention their impact on our planet's climate. The dilemma is at the center of a broader philosophical debate concerning the Anthropocene, the current geological era in which human activities significantly impact the Earth, raising pressing questions about the ethics of responsibility to future generations.

Addressing such trade-offs requires a sophisticated understanding of energy efficiency, sustainable design, materials science, and a collective commitment to reducing energy consumption and adopting renewable sources. For example, transitioning to renewable energy is crucial to lowering the technology industry's carbon footprint. In addition, innovation in sustainable materials and advanced recycling technologies opens up new possibilities for reducing the environmental impact of electronic devices.

The challenge is to develop and use such technologies to maximize their benefits and minimize their thermodynamic costs.

This task is not easy, involving difficult choices and complex trade-offs, but it is an essential part of our ongoing struggle to make the most of the energy we have. Ecological economics theories and the concept of degrowth echo this theme, calling for rethinking our development model in terms of greater equity and sustainability.

In our efforts to move forward, we must, therefore, carefully navigate between the immediately apparent demands of innovation and the less visible but equally tangible consequences of our energy and environmental impacts. All this requires a holistic view that integrates technology, ethics, and ecology, an approach that philosophers such as Hans Jonas have explored, emphasizing the need for an ethics of responsibility in a technologically advanced age. Only then can we aspire to a future in which technological progress and environmental sustainability go hand in hand.

8.8 The future of technology

As we peer into the future of technology, the role of thermodynamics becomes more and more apparent. The desire for greater efficiency, speed, and more powerful devices pushes us ever closer to the boundaries of what is thermodynamically possible. At the same time, the environmental and energy implications of technological advances require us to pay close attention to the thermodynamic costs of our choices. The debate reflects the growing awareness of the carbon footprint of digital technology, with recent studies showing that the ICT (Information and Communication Technologies) industry is responsible for about 2 to 3 percent of global greenhouse gas emissions, a percentage comparable to that of the aviation industry.

In this context, *Koomey's law*, which predicts more energy-efficient computers over time, clashes with the physical limits of energy and matter, a battle outlined in discussions of the limitations of *Moore's law*. While the latter has successfully predicted the doubling of transistor density in integrated circuits every two years, studies such as those published in *Nature* suggest that we are reaching the point where economic and thermodynamic considerations could end this trend.

The *Landauer limit*, which establishes a minimum limit to the heat generated by each bit of information erased during computation, is an example of a fundamental thermodynamic barrier that influences the future evolution of technology. This principle, related to the irreversible nature of computation, has been confirmed experimentally in studies such as those published in the journal *Science*, underscoring how fundamental limits in physics affect the performance of our computers.

Looking ahead, one of the major trends is the continued miniaturization of electronic components. As we pack more and more transistors onto a silicon chip over time, we face increasing challenges in terms of managing heat dissipation, an essential aspect of thermodynamics. Quantum computing, which promises to revolutionize computing by exploiting quantum mechanics to perform computations much more efficiently than classical computers, is also grappling with significant thermodynamic challenges, including the need for extremely low operating temperatures. Progress in this field is constantly reported in scientific publications such as *Nature Physics*, highlighting the race to achieve quantum supremacy.

Quantum cooling systems and superconducting materials are at the forefront of these challenges. They try to keep qubits stable and functional while minimizing quantum decoherence caused by heat. Research in this area is intense, and developments are reported in academic journals such as *Advanced Materials*, which document how new materials and designs can contribute to the scalability of quantum computing.

The energy sector is witnessing intense research on new technologies, such as perovskite solar cells and smart grids, to effectively integrate renewable energy into our existing infrastructure. The former, for example, have been highlighted for their efficiency and low-cost potential in publications such as *Science Advances*. At the same time, studies published in IEEE Transactions on Smart Grid have recognized the latter's ability to optimize energy use.

In the energy field, the push toward renewable sources such as wind, solar, and hydropower represents a significant shift in our thermodynamic strategy. By tapping into the energy flows of the Earth and the Sun, the technologies offer the promise of sustainable energy use that minimizes the increase in entropy. According to the International Energy Agency, renewables are set to account for nearly 30 percent of global electricity generation by 2024, underscoring their growing role in addressing climate change. A shift toward cleaner and more sustainable sources reflects a paradigm shift inspired by growing awareness of the impact of human activities on the global environment. At the same time, developing more efficient batteries and energy storage systems will be critical to managing the intermittent nature of many renewable energy sources.

Innovation in lithium-air batteries and fuel cells is moving closer to the *perfect battery*, with higher energy density, long life, and low environmental impact. Recent studies published in scientific journals such as *Nature Energy* and the *Journal of Power Sources* highlight advances in battery technology, suggesting that such innovations could revolutionize energy storage, making it more efficient and affordable. These developments represent a milestone in the search for sustainable solutions, offering new perspectives for the global energy balance.

With the advent of 5G networks and beyond, we are entering a new era of connectivity that promises to increase further energy demand and the challenges in managing the associated entropy. The fifth generation of mobile telecommunication technologies, with its increased speed and capacity, enables an unprecedented amount of data transfer, which, as recent studies point out, could increase the energy consumption of network infrastructure by up to 170 percent by 2026.

In the digital realm, we can expect the Internet of Things (IoT) to continue to expand, with more and more devices-from home appliances to industrial equipment-connecting to the Internet. The proliferation of connected devices offers many benefits, but it also increases the overall energy consumption of our digital infrastructure. According to research institute Gartner, the growth of IoT is expected to exceed 25 billion devices by 2025; this inevitably leads to and necessitates the need for consideration of how devices can be designed for energy efficiency from cradle to grave, incorporating sustainable design principles and energy recharging technologies such as ambient energy harvesting.

Exponential IoT growth implies a significant increase in digital entropy, which must be managed through design and standardization to avoid unsustainable energy overload. Adopting low-power communication protocols and integrating artificial intelligence solutions to optimize energy use are examples of how technology can address these challenges, promoting a more conscious and responsible approach to digital innovation.

Similarly, advances in artificial intelligence (AI) promise to revolutionize many aspects of our lives, from health care and transportation to entertainment and communication. However, energy consumption for training large machine learning models is a significant concern. Research on energy-efficient AI and simplified machine learning seeks to reduce the energy footprint. At the same time, lightweight algorithms and optimized neural network architectures promise to bring the power of AI to energy-constrained devices.

Understanding the thermodynamic implications of our technology choices can help us choose a path to a future that maximizes technology's benefits while minimizing energy costs and environmental impact.

Reflecting on Albert Einstein's words, "The mind that opens to a new idea will never return to its original size," we are invited to explore the frontiers of science and technology with an ecological and sustainable consciousness.

Navigating the future requires a delicate balance between exploration and conservation, innovation and sustainability, as we seek to extend the frontiers of technology without compromising our planet's resources. French philosopher Edgar Morin recalls the importance of a holistic approach to knowledge, emphasizing that the complexity of contemporary challenges requires interdisciplinary understanding and global cooperation. Integrating ethical principles into technological design and development emerges as an imperative, directing research toward solutions that respect the balance of our ecosystem and promote collective well-being.

8.9 Space exploration

The final frontier, as it is often called, represents a unique and fascinating thermodynamic challenge. Space exploration has always been a high-stakes environment where every ounce of fuel and every watt of power is of critical importance. The race toward space exploration, which began in the 20th century with pioneers like Yuri Gagarin and historic missions like Apollo 11, continues to evolve, pushing us beyond the known boundaries of the universe. Step by step, humanity is pushing further into the cosmos, involving a progressive need to confront thermodynamic limits in new and complex ways, a journey that tests our technological ingenuity and our resilience and adaptability as a species.

Traveling through the space vacuum poses unique problems, such as cooling spacecraft. The absence of an atmospheric medium makes heat transfer via conduction or convection impossible, leaving only radiation as the mode of heat transfer. This principle was highlighted during the Apollo missions, where heat management was a critical consideration for the success of lunar missions. Thermal radiation became the only effective mechanism for dissipating the accumulated heat from electronic systems and engines, requiring innovative thermal engineering solutions.

Propulsion is the most obvious aspect of space exploration and involves thermodynamics. Conventional chemical rockets are remarkably inefficient, converting only a tiny fraction of the fuel's potential energy into the kinetic energy of forward motion. Even state-of-the-art propulsion technologies, such as ion engines or nuclear propulsion, still operate within the limits of the laws of thermodynamics. The relentless quest to overcome the limits of traditional chemical propulsion has led to the testing of ion thrusters on missions such as NASA's Dawn, which explored Vesta and Ceres, demonstrating the potential for innovative technologies to extend our reach into deep space.

The use of nuclear reactors for space propulsion, known as nuclear thermal propulsion, could significantly improve the energy efficiency of prolonged space travel, a prospect studied by NASA and other space agencies. Nuclear thermal propulsion, with its potential to provide a higher thrust-to-weight ratio than conventional engines, could dramatically reduce travel times for missions to Mars and beyond, marking a significant step toward space colonization.

But propulsion is only the beginning. Life support systems in spacecraft and other planets or moons also have significant thermodynamic considerations. Such systems must provide breathable air, potable water, and comfortable temperatures for astronauts, all in an enclosed environment where waste heat, carbon dioxide, and other byproducts cannot simply be dumped into the environment. Regenerative life support systems, which recycle waste products into usable resources, provide an ingenious challenge to the trend of increasing entropy by creating pockets of order in the otherwise inhospitable environment of space. The International Space Station (ISS) serves as an on-orbit laboratory for testing these technologies, including water recycling and life support systems that minimize the loss of life resources.

Managing energy and entropy in closed systems such as space stations and future planetary colonies is crucial, where the principles of thermodynamics must be applied so that resource recycling and reuse systems are integrated by design. The holistic approach to sustainability in space reflects the closing theories of ecological cycles on Earth and is critical to ensuring the long-term viability of space colonies.

Communication across the great distances of space also involves thermodynamics. Radio signals scatter and weaken as they travel, and their transmission requires energy. The search for extraterrestrial intelligence (SETI) must confront thermodynamic realities as it scans the skies for feeble signals when they reach us. Projects such as the Green Bank Telescope and the Very Large Array are examples of how scientists overcome such challenges, using advanced technologies to amplify and analyze signals in search of extraterrestrial communications.

Technologies such as adaptive optics and space telescopes are developed to maximize the collection of faint signals in low-entropy environments, making the best use of atmospheric transparency windows and minimizing terrestrial interference. These instruments, including the Hubble Space Telescope and the James Webb Space Telescope, improve our ability to observe the universe and understand its fundamental laws. This leads to discoveries in physics and astronomy that challenge our understanding of entropy and thermodynamics.

Looking further into the future, concepts such as terraforming other planets or building Dyson spheres to capture a star's entire energy output involve astronomical thermodynamics. The idea of a Dyson sphere, first proposed by physicist Freeman Dyson in 1960, requires an engineering concept so advanced that it defies our current scientific and technological limits. Grandiose projects, if ever made a reality, would represent some of the most daring acts of thermodynamic defiance in the universe, equivalent to a rewrite of planetary and stellar physics as we know it.

Terraforming Mars, for example, would require an immense input of energy and resources to modify an entire planet capable of sustaining human life. This task could require the release of greenhouse gases to warm the planet or the creation of artificial magnetospheres to protect the atmosphere. According to a 2018 study published in *Nature Astronomy*, the current inventory of greenhouse gases on Mars must be increased to provide significant warming through the greenhouse effect. That task may also require using as-yet undeveloped technologies, such as synthesizing powerful greenhouse gases or large space mirrors to reflect more sunlight onto the Martian surface.

As we continue our journey through the cosmos, the laws of thermodynamics will be our constant companions. They will guide technologies, challenge our ambitions, and offer profound insights into the nature of the universe we seek to explore. Entropy, a measure of disorder in a system, not only defines the limits of usable energy but also confronts us with the mystery of temporal asymmetry and the arrow of time. This contemplation attracts the attention of physicists and philosophers alike.

Our advance to the stars will require a balance between daring innovation and scientific rigor. Each step into deep space must be precisely calculated to ensure survival and success. Emerging theories, such as string theory or *quantum gravity loops*, could define the future of space exploration. These theories would provide new models for understanding energy and matter on cosmic scales.

8.10 Technology and the Second Law: a final frontier

In its relentless pursuit of progress, technology stands at the edge of the second law of thermodynamics. It represents humanity's bold challenge to the universe's propensity for disorder, the way to assert some control over entropy. A conflict that is not only central to the present but will also dictate the trajectory of the future. A technology shaped as a manifestation of the innate human tendency toward innovation and adaptation, a resistance against the relentless tide toward disorder symbolized by entropy.

The concept of *negative entropy*, proposed by physicist Erwin Schrödinger in his essay *"What is Life?"*, becomes relevant. Schrödinger suggests that living organisms and the technologies they develop distill order from their environment to maintain or increase complexity. Schrödinger describes life as an island of order in a sea of disorder. Today, we can see technology as an extension of that concept, where machines, algorithms, and systems seek to create and maintain order in a universe that naturally tends toward chaos.

In many ways, modern technology is a story of increasing efficiency: squeezing more work from less energy and creating more order from less. From the heat engines of the Industrial Revolution, limited by the theoretical limit of the Carnot cycle, to today's ultra-efficient quantum computers that flirt with the limits of the Landauer principle, we have progressively learned to use our energy resources more. A path that maps our ingenuity and our aspiration to transcend the boundaries imposed by nature, attempting to write a new chapter in the physical laws that govern our world.

Recent processor developments have increased energy efficiency to unprecedented levels. Two-dimensional materials such as graphene promise to further reduce waste heat and push efficiency far beyond current limits. With its incredible conductivity and almost miraculous structure, graphene represents the frontier of nanotechnology and could be the key to unlocking new regimes of low power consumption and high efficiency.

While we have become more efficient at the micro level, we are generating more entropy at the macro level than ever. This paradox is reflected in *Kardashev's law*, which ranks civilizations by the amount of energy they can use. Suppose we aspire to become a Type II civilization, mastering the energy of an entire star. In that case, we would also need to think about managing and minimizing the entropy we create, facing sustainability challenges on a cosmic scale.

Paradox highlighted in models of exponential growth in energy consumption, such as those discussed in *Jevons' Paradox*, which suggest that increasing efficiency leads to an overall increase in resource consumption. The paradox, first formulated by economic historian William Stanley Jevons in the 19th century, notes that improvements in energy efficiency can lead to a reduction in cost per unit of consumption, which in turn can stimulate an increase in total demand for that resource rather than the expected reduction. Our constant drive for improvement and growth, fueled by technology, may seem like a challenge to the second law. However, we must come to terms with the knowledge that the second law will always remain undefeated. The energy used by our technology will eventually increase the universe's entropy. Therefore, sustainability in the face of the second law becomes a crucial technological frontier. This is where the principles of thermodynamics and sustainability must be integrated, not only in energy production but in all stages of technological product creation, use, and life cycle.

Addressing such a frontier means embracing concepts such as thermodynamic efficiency within planetary limits and developing technologies that harmonize with the laws of nature rather than trying to subdue them. The idea of *degrowth* or *green growth* is an approach that aims to balance technological progress with the regenerative capacity of natural ecosystems and the demands of sustainable energy consumption.

This does not mean that we are doomed to stalemate. On the contrary, it pushes us to be more creative and innovative. Recycling, green energy, and circular economy models are all human strategies to cope with thermodynamic realities. Future technology must be more efficient and holistic, considering its impacts from a broad thermodynamic perspective. Incorporating systems thinking and life cycle analysis into the design of technologies can help us develop systems that minimize entropy during their use and consider the entire span of their existence from production to disposal. One example is the so-called *cradle-to-*

cradle approach *(from cradle to cradle)*, which considers each product as a temporary service that can be fully reused or re-enter the natural cycle without generating waste once it has served its purpose. Design philosophies and emerging technologies can chart a technology course that is in harmony with our planet and its delicate thermodynamic balances, pushing us toward a more sustainable and responsible future.

Notes and bibliographical insights in Chapter 8:

1. Martin Heidegger, *The Question Concerning Technology* (1954). Heidegger discusses the concept of technology as a means of revelation, which transforms the natural world into resources available to humankind, thus exploring the role of humans as both the subject and object of technology.

2. Garrett Hardin, *The Tragedy of the Commons* (1968). Hardin examines how communal management of limited resources often leads to long-term negative consequences, a principle that applies to fossil fuel use and sustainability challenges.

3. Karl Popper, *The Logic of Scientific Discovery* (1934). In this book, Popper introduces the concept of falsifiability in the sciences, which can be linked to technological innovation's experimental and iterative nature.

4. Stuart Kauffman, *At Home in the Universe* (1995). Kauffman discusses the adjacent possible and emerging complexity in biological and technological systems, offering a perspective on the progressive complexity of technology.

5. Claude Shannon, *A Mathematical Theory of Communication* (1948). Shannon founded the theory of information, linking the concept of entropy to information processing and transmission, which was fundamental to the computer science era.

6. Leonard Susskind, *The Black Hole War* (2008). Through the black hole debate, Susskind explores the link between information and thermodynamics, illuminating the relationship between physics and information technology.

7. Jared Diamond, *Collapse: How Societies Choose to Fail or Succeed* (2005). Diamond extends his analysis of past societies to modern ones, examining how technological and energy choices affect long-term sustainability.

8. Nicholas Georgescu-Roegen, *The Entropy Law and the Economic Process* (1971). Georgescu-Roegen applies the principles of thermodynamics to economics, emphasizing how entropy affects technologies and growth limits.

9. Thomas Hughes, *Networks of Power: Electrification in Western Society, 1880-1930* (1983). Hughes examines how technological innovations such as electrification transformed societies and power structures, reflecting on the role of energy in technological progress.

10. Freeman Dyson *Imagined Worlds* (1997). Dyson reflects on how future technologies could transform society, touching on concepts of thermodynamics and the limits of physical laws in technological evolution.

Chapter 9

FUTURE:
LIFE, THE UNIVERSE, AND ENTROPY

From the Big Bang to Consciousness

Man is only a reed,
the weakest in nature,
but he is a thinking reed.
There is no need for the whole universe to take up arms to crush him:
a vapour, a drop of water is enough to kill him.
But even if the universe were to crush him,
man would still be nobler than his slayer,
because he knows that he is dying
and the advantage the universe has over him.
The universe knows none of this.

PASCAL BLAISE
Pensées, (1670)

From the Big Bang to Consciousness

9.1 The Distant Future

The future of our universe is inextricably linked to the laws of thermodynamics. While previous chapters delved into the details of how these laws shape life, society, and technology, we have yet to consider the final and perhaps most profound implication: the distant future of the universe itself. The *Big Freeze* theory, the *Big Rip*, or the *Big Crunch* are scenarios theorized by astrophysicists that depend entirely on the laws of thermodynamics and the universe's fundamental properties of matter and energy.

In such a context, the concept of the *end of time* becomes fundamental, suggesting a future in which time flow may become less meaningful due to thermodynamic uniformity. Some theorists suggest that in a universe where entropy has reached its maximum, time, as we perceive it, may lose its directionality. This concept challenges our understanding of physics and reality itself.

According to the second law of thermodynamics, in any closed system, entropy – the disorder – will always increase over time. As far as we know, our universe is the ultimate closed system. So what does the future hold, given the relentless march toward disorder? Modern cosmology, armed with detailed observations and advanced mathematical models, seeks to predict the long-term evolution of the universe, considering the ultimate fate of everything from galaxies and stars down to individual particles.

To answer these questions, special attention must be paid to stars, as they can be understood as the most visible manifestations of energy conversion in our universe. They can fuse lighter elements into heavier ones and, in the process, release the light and heat that fuel life on Earth. But stars do not shine forever. They exhaust their nuclear fuel at the end of their life cycle, leaving behind dark remnants such as white dwarfs, neutron stars, and black holes. Understanding *stellar death* is essential for cosmology and raises fundamental questions about the nature of time and existence.

The latest celestial objects, particularly black holes, pose intriguing questions about their role in the thermodynamic future of the universe, given their intense gravity and the phenomenon of Hawking evaporation, which suggests how they might have a mechanism through which they succeed in losing mass and energy.

Hawking evaporation, a theoretical perspective proposed by Stephen Hawking in the 1970s, presents a possible end for black holes that challenge our notions of eternity and immutability. If black holes evaporate, they could be part of the thermodynamic cycle of the universe, contributing to its entropy and long-term evolution. This would imply that the universe is destined to become a uniformly cold and dark place dominated by radiation or that it might have more complex dynamics than we currently imagine.

From a thermodynamic perspective, each final state represents an increase in entropy. The process of stellar evolution can thus be seen as a transformation from states of lower entropy (highly ordered and concentrated energy) to states of higher entropy (less ordered and dispersed energy). The transition from the fiery core of a star to the cold expanses of the post-stellar cosmos is a microcosm of the general direction that thermodynamics imposes on the universe. It is a narrative that extends from billions of years in the past to the inconceivably distant future.

Hypothetical processes that challenge our current understanding, such as the annihilation of dark matter or the emergence of new physical laws that come into play on extraordinarily long-time scales, may also affect the distant future. Dark matter, which constitutes about 27 percent of the universe's total mass, remains one of the most persistent mysteries of modern physics. Its potential annihilation or other undiscovered interactions could alter the pathways of cosmic evolution in ways we can only theorize about.

If we extend the same line of thinking to the entire universe, we arrive at a future scenario known as *heat death*. In this state, all the energy in the universe has been evenly distributed, and no more work can be done. A definite increase in entropy: a state of maximum disorder in which all differences in energy, and therefore all potential for change, have been eliminated. It is a concept that embodies the depth of thermodynamic *nothingness*, where the universe would continue to expand forever but without the possibility of generating new forms of complexity or life.

Some theorists have contemplated the idea of *bubble universes*, in which new universes could be born inside black holes, escaping the thermal death of the host universe through a process of cosmic reproduction. Such scenarios, derived from interpretations of physical theories such as inflationary cosmology, offer a glimmer of continuity or even rebirth beyond our universe's seemingly final thermodynamic decline.

Yet such a bleak scenario may not be the end of the story. From the birth of life to its constant adaptation and evolution, the universe has demonstrated an extraordinary capacity for ingenuity. What place might life and intelligence have in the distant future, driven by thermodynamics? Our investigation may reveal that intelligence, once emerging, may find unexpected ways to persevere or even thrive.

Speculations about *post-biological life* and advanced civilizations capable of manipulating cosmic structures suggest that understanding and managing entropy could be the center of long-term survival strategies in the vast universe stage. Sufficiently advanced civilizations could develop means to control or reduce entropy or even harness the very nature of cosmic expansion and cooling. Such civilizations would come to produce their own energy and resources in ways that currently escape our understanding, potentially perpetuating life and intelligence far beyond the limits predicted by classical thermodynamics.

9.2 The heat death of the Universe

The term *heat death* may conjure up the image of a fiery, chaotic end to the universe, but the reality is precisely the opposite. Death by heat refers to a state of *maximum entropy* in which all energy in the universe has been evenly distributed, and all processes that could concentrate energy in one place have run their course. Such a universe would be cold, dark, and lifeless. Death by heat is the silent, still twilight of the universe, an eternity of stillness after the last star has ceased to shine.

In physics, the scenario of universal thermal equilibrium, where the temperature difference, and thus the potential for work and energy, no longer exists, is known as the *Great Cold Death*. This fate embodies the extreme realization of thermodynamics on a cosmic scale. In this scenario, the universe reaches a state of thermal indifference, with heat no longer flowing because it has the same temperature everywhere.

A somber image, a logical consequence of the second law of thermodynamics and the concept of entropy. As we discussed earlier, in any closed system-including our universe-total entropy tends to increase over time. Energy tends to distribute evenly, leaving the universe in a perfect but inert equilibrium.

But what does this mean in practice? Today, the universe is far from a heat-dead state. Energy is unevenly distributed, with hot stars shining in the cold vacuum of space and complex structures such as galaxies, stars, and life forms creating pockets of order in the cosmic chaos. The disorder, the incredible variety, and the beauty, from the dance of galaxies to the succession of seasons on Earth, are all the result of entropy *not yet* balanced, a song still in full song before it settles into ultimate silence.

Current theories suggest the universe will take trillions of years to approach a heat-dead state. This is an unimaginable time frame in which trillions of stars will be born and die, galaxies will collide and merge, and black holes will swallow matter before evaporating away, as proposed by Hawking radiation. Eventually, even black holes may vanish, leaving only a faint luminescence of cosmic background heat as an echo of the universe that once was.

Current cosmological theories, such as the *Big Freeze* model, imply that the universe's accelerated expansion will eventually overcome all forces holding cosmic structures together, leading to the final isolation of all energy sources. The gradual receding of galaxies and galaxy clusters from one another is an observed and measured phenomenon known as *Hubble's law*. This law reflects the dynamic character of a constantly changing universe and the isolated fate that awaits each celestial formation.

Over time, stars will fade, galaxies will recede, and even black holes – the last reservoirs of matter and energy – will slowly evaporate through a process called *Hawking radiation*. In the distant future, there may be no more structure, just a thin soup of evenly distributed particles in an ever-expanding universe. Hawking radiation, a prediction of theoretical physics that still awaits observational confirmation, represents one of the few ways the universe could shed its last energy content, ending the life-and-death cycle of its most massive structures.

Some physicists have suggested that processes such as Hawking radiation could take such unimaginably long times, on the order of 10^{100} years or more, to complete the evaporation process of black holes. Such time periods defy any comparison with the human time scale and underscore the immensity not only of space but also of cosmic time. Our existence is only a fleeting instant compared to such epochs, yet it is in this brief moment that the universe becomes self-aware through life and intelligence.

It is still entirely speculative and depends on our current, incomplete understanding of physics. However, the concept of heat death still provides a helpful framework for considering the ultimate fate of the universe and the inexorable march of entropy. It reminds us that although the laws of physics are immutable and universal, our understanding of them constantly evolves, subject to revision and expansion as our technology and wisdom expand.

Yet even in the face of such an inevitable outcome, life, as we have seen, has been a stalwart soldier in the battle against entropy. Could life and intelligence, in some as-yet-ununderstood form, find a way to persist or even thrive? Could consciousness itself have a role in the cosmos beyond mere survival? Questions that prompt us to reflect not only on our place in the universe but also on the potential for intelligence to influence or interpret the laws of nature in new and surprising ways.

9.3 The struggle of life against the inevitable

Against the cosmic backdrop of starlight and the expanding space-time canvas, life presents a strange but fascinating counterpoint to the entropic march of the Universe. For billions of years, through the process of evolution, life has demonstrated its remarkable tenacity and extraordinary ability to harness energy to build complex structures and processes, always seeming to defy the inexorable pull of entropy. Biological adaptation presents itself as a continuous dialogue between living organisms and the challenges posed by their environment, a dynamic exemplified by evolutionary history. Charles Darwin's observation of evolution reveals how life optimizes the use of energy to maximize survival and reproduction. It is a narrative that unfolds on epic time scales, in which every adaptation, every mutation, and every life cycle is a note of resilience and innovation.

A struggle at once tragic and sublime where life, as we understand it, cannot exist in a state of maximum entropy. Complex processes such as metabolism, replication, and cognition require energy gradients: they feed on pockets of low entropy in the universe. The heat death of the universe, as described in the previous section, seems to mark an inevitable fate for life. Yet, just as the stars shine brightly against the black velvet of the night sky, life takes on extraordinary meaning and beauty against the backcloth of growing entropy.

Nevertheless, *emergent complexity theory* suggests that systems of great complexity can arise from simple rules and interactions at the local level, implying that, under the right conditions, life would be able to emerge or persist in ways we do not currently fully understand, exploiting phenomena such as symbiosis, cooperation, and innovation. In a universe where heat dies, life would still be able to find niches of energy and order from which to extract its essence, just as extremophiles on Earth find ways to thrive in environments that would be inhospitable to other life forms. There may be a kind of life we cannot yet imagine, operating on radically different temporal and spatial scales than terrestrial life, a form of existence that transcends our current biological paradigms. In some exotic form, a potential persistence of life offers a spark of hope and wonder in an otherwise austere, inescapable context. Yet the story does not necessarily end there. Life's struggle against entropy is not just a story of survival against the odds. It is a pivotal testimony to the ingenious ways nature has found to harness energy and use it to create complexity and order. It is the story of how genetic information and natural selection have shaped life forms capable of exploiting the environment to perpetuate themselves, from photosynthesis to animal migrations driven by seasonal changes.

Think of life as an undertow in a stream, a temporary local reversal of the general trend of increasing entropy. Even as the universe cools and expands, life keeps going, constantly adapting, finding new ways to survive and thrive each time. An adaptation that is not a mere accident but the result of billions of years of evolution, a demonstration of the principle that life finds a way.

Despite the seemingly ominous prognosis provided by the second law of thermodynamics, life has been incredibly resilient and adaptive, and we have no reason to believe that it will not continue to be so.

Considerations of the role of consciousness and collective intelligence in the universe lead us to ponder hypotheses such as the *Great Filter* or *Cosmic Transition*, which postulate critical points in the development of advanced civilizations capable of overcoming the challenges imposed by thermodynamics and leaving a lasting imprint on the universe. The most significant challenges facing intelligent life are not the external environment but its ability to understand, adapt to, and manipulate the environment at ever deeper levels. The actual test of an advanced civilization: not just surviving in the short term but successfully navigating through thermodynamic and cosmological hazards in the long term, harnessing knowledge and intelligence to forge a destiny that transcends natural limits.

9.4 The future of life: a thermodynamic forecast

Projecting the future of life on the basis of thermodynamics is a daring task. As a consummate manipulator of energy gradients, life has demonstrated incredible resilience and adaptability. This marked ability to survive and transform suggests that life may persist despite the challenges posed by thermodynamics, and new ways may flourish in previously inhospitable environments.

The *theory of panspermia*, for example, suggests that life might be much more ubiquitous in the universe than is generally believed, meaning it is capable of traveling between planets and even stars. If this were true, it would offer unprecedented potential for life to spread and adapt to changing thermodynamic conditions on a cosmic scale. It is a perspective that broadens our concept of life, suggesting that it might not be confined to the limits of our planet but instead be an intrinsic feature of the universe, ready to exploit any habitable niche.

In the short term – relative to cosmic timescales, of course – the next few billion years present opportunities for life to continue to thrive, at least on Earth. In a gradual process, our Sun will age, becoming hotter and increasing the amount of energy available to Earth's biosphere. Life will have to adapt to such changing conditions. Still, given the immense variety of life forms on our planet and the robustness of evolutionary processes, it is reasonable to expect that life will find a way to do so. This transition period could witness the emergence of new life forms adapted to take advantage of rising temperatures and energy resources.

According to the expanding H*Z* (*habitable* zone) *model*, the growth of our Sun's habitable zone would result in the proliferation of life in places that are too cold today, such as the icy moons of Jupiter and Saturn, where the presence of liquid oceans beneath their surfaces cannot be ruled out. Such an expansion of the habitable zone offers a window of opportunity for extraterrestrial life, perhaps even within our own Solar System, demonstrating that the dynamics of the universe can, at certain times, favor the spread of life rather than hinder it.

In the medium term, Earth's future will look much less rosy when the Sun exhausts its nuclear fuel and expands into a red giant. This does not necessarily mean the end of life in the Solar System. Moons around gas giants such as Jupiter and Saturn, expected to warm, could become hospitable for life. With their potential underground oceans, such celestial bodies would become the new hotbeds of extraterrestrial biology. In a long-term thermodynamic scenario, the future of life could depend on its ability to find refuge in new habitable niches, continuing the cycle of adaptation and innovation that has characterized its history on Earth. As the universe continues its march toward entropic equilibrium, life, in its most resilient and adaptive form, may yet find ways to assert its presence, even in the context of far-reaching cosmic changes.

Recent advances in telescope technology, such as the James Webb Space Telescope, are facilitating the discovery of more clues about possible extraterrestrial biospheres, thus extending our horizon on the likely development of life beyond our home planet. Advanced instruments that open an unprecedented window on the universe, allowing us to peer into planets located hundreds or thousands of light-years away and investigate their atmospheres for signs of habitable conditions or even bioformations. A perspective that transforms the search for extraterrestrial life from a philosophical speculation to a concrete scientific possibility.

Beyond our Solar System, the galaxy abounds with stars, many of which will continue to burn long after our Sun is gone. It cannot be ruled out that some of these star systems have planets within them with conditions favorable to life. Local energetic conditions would dictate the evolution of life on one of these planets, just as has occurred on Earth. The diversity of stars and planetary systems in our galaxy and beyond suggests that the opportunities for life to emerge and evolve could be many and varied, offering a myriad of scenarios in which life could adapt and thrive.

The concept of the *cosmic Goldilocks zone* suggests the presence of regions of the universe where thermodynamic conditions are optimal for life. If we could identify such regions, we might find even more diverse and complex ecosystems than those on Earth in a search that extends the habitable zone principle beyond the boundaries of our Solar System, proposing the idea that life's ability to sustain itself depends on a combination of cosmic factors that go beyond mere distance from a star.

In the long term, as the universe expands and cools, the energy gradients that fuel life will become increasingly scarce. Even in this distant epoch, pockets of relatively low entropy could persist for long periods. If life could harness these islands of order in the cosmic sea of entropy, it could persist into the future long after the last stars have died out. The ultimate adaptive capacity underscores the incredible potential of life to navigate and exploit the cosmic environment, no matter how extreme it may become. One could imagine life clinging in microcosmic enclaves, such as around primordial black holes, where the laws of physics would allow atypical sources of energy to be harnessed, exotic niches that could provide shelter for life in forms we can begin to imagine, sustained by physical processes yet to be fully understood.

Although these scenarios are speculative, they underscore a crucial insight: the future of life is intricately linked to the future of energy and entropy in the universe. As we explore the implications of this connection, we will see that it leads to profound and far-reaching questions about our place in the cosmos. The understanding that life, at its most fundamental, is intertwined with the dynamics of the universe itself invites us to reflect on our responsibility as an intelligent species: to preserve and promote life not only on Earth but, perhaps one day, across the stars.

9.5 The role of intelligence in the future of the universe

In contemplating the future of life, we must also consider intelligence's role. From a thermodynamic perspective, the appearance of intelligence on Earth represents a new way for life to manage and exploit energy gradients. Intelligent beings – humans, specifically – can understand and manipulate their environment in ways that no other species can, affecting the planet's thermodynamic balance. This unique ability to actively influence one's environment opens up new avenues for managing entropy and sustaining life under increasingly complex and variable conditions. Reflection on the role of intelligence expands beyond mere survival; it embraces Robin Hanson's *grand filter* theory, which postulates that intelligence is a rare and challenging threshold to cross in cosmic evolution. If so, humanity could be among the pioneers in a universe still waiting to be awakened by intelligence. Such a perspective raises profound questions about the rarity of intelligent life and the responsibilities of overcoming such a cosmic filter, including the potential custodianship of livable environments or the dissemination of life beyond Earth.

Our increasing understanding and control of the physical world will soon enable us to harness energy more efficiently and, perhaps, discover new forms of energy. In a world where energy use is optimized, we may see a decrease in entropy production, at least on a local scale. This has implications for our planet's sustainability and our ability to project life and intelligence beyond Earth's boundaries.

Everything could be accelerated by the development of advanced artificial intelligence that would ideally be able to find energy solutions that elude us. The convergence of AI and nanotechnology would lead to revolutionary ways of capturing, storing, and using energy with technologies that could improve energy efficiency but also allow us to manipulate the environment on previously unimaginable scales, paving the way for new strategies to counter entropy and sustain life in increasingly hostile environments.

It cannot be ruled out, then, that natural and artificial intelligence would play a central role in the universe's future, making us not mere observers but active players capable of shaping our destiny and that of other living systems. The interplay between intelligence and thermodynamics suggests a future in which life and consciousness continue to flourish, despite the challenges posed by entropy, through continued innovation and the exploration of new energy frontiers. Thus, as the universe expands and cools, the flame of intelligence may continue to burn brightly, perhaps even discovering new ways to challenge or circumvent the thermodynamic laws we now see as inescapable limits.

In the medium term, the survival of life in our solar system could depend on our (or artificial intelligence systems we create) ability to migrate to other planets, moons, or even artificial habitats. The challenges of space colonization are formidable, but they are essentially energy problems and, therefore, potentially solvable through innovation and intelligent engineering. Growing human mastery of space technology opens up the possibility of overcoming obstacles that once seemed insurmountable, transforming space exploration and colonization from science fiction dreams to concrete engineering projects.

Ion propulsion, solar sail, and space elevator concepts are examples of how we might overcome energy barriers to space colonization. Each approach offers a unique way to reduce energy costs and increase the efficiency of interplanetary and interstellar travel, making human expansion into space more feasible. While these technologies are still in the development or refinement stage, they represent critical steps toward achieving a sustainable, long-term human presence beyond Earth.

In the very long term, intelligence could become crucial to the continuation of life in an increasingly hostile universe. Suppose life is to survive after the stars are depleted. In that case, it will be necessary to rely on intelligent manipulation of energy at a fundamental level, potentially involving processes and technologies that are currently beyond our understanding – for example, perhaps exploiting quantum properties of the universe or harnessing energy processes around black holes, as Freeman Dyson suggests.

We may also have to consider Dyson's idea of life and intelligence around black holes, where extreme gravitational energies would provide new thermodynamic opportunities by offering a lifeline for life in a cosmic age of scarce energy resources and allowing life to exploit the last foci of energy in the dying universe.

Moreover, we should consider the possibility that our form of biological intelligence is only one point in a broad spectrum. The future could see the rise of post-biological intelligence, either in the form of advanced AI or due to fusion with our technology. Such entities could possess thermodynamic efficiencies far beyond what is possible for us, opening up new possibilities for the future of life. Such bits of intelligence might be able to navigate the later stages of the universe, finding ways to persist and even thrive.

Like the life forms suggested by Nick Bostrom, artificial intelligence can dominate the cosmic information age, where information processing becomes the primary energy user in the universe. In a scenario where the ability to process and store information could become the ultimate feature of advanced life, artificial intelligence is crucial in managing scarce but optimized energy resources for maximum thermodynamic effect.

Pushing the boundaries of our understanding, we might discover that life still has tricks to unveil, reflecting on how intelligence, whether in biological or post-biological form, would be able to shape the future of existence in an ever-evolving universe. The fusion of human ingenuity with the potential of advanced technology and artificial intelligence would thus succeed in ensuring the survival of life and opening up incredible new cosmic horizons for it.

9.6 The possibility of life elsewhere in the Universe.

Given the vastness of the universe, the existence of life elsewhere seems presumptive. However, considering stars as numerous as grains of sand on Earth's shores opens the mind to infinite possibilities of worlds and life forms. This sense of wonderment is reinforced by understanding how life, in its essence, is a manifestation of physics and chemistry, universal processes impossible to confine within the limits of our planet.

The Drake equation, formulated in the 1960s by scientist Frank Drake, seeks to estimate the number of communicating civilizations in our galaxy. While steeped in uncertainty, it opens the door to the possibility of intelligent life beyond Earth. It is a pioneering attempt to quantify the unknown that reflects a deep desire to understand our place in the universe and suggests that we are not alone in vast cosmic space.

If we accept that life is a natural response to the presence of energy gradients as a consequence of thermodynamics, it should occur wherever conditions allow. The notion of life emerging as a way to dissipate energy more efficiently aligns biology with the fundamental laws of physics, suggesting that life might inevitably take hold wherever suitable conditions exist.

For example, the recent detection of phosphine in Venus's atmosphere suggests that life can thrive even in extreme environments that are very different from Earth's. A detection that raises the hypothesis that life is not only possible in Earth-like conditions but can adapt to a much more comprehensive range of environments than previously imagined.

The massive number of stars, each potentially with its own planetary system, implies that such conditions could exist in countless places. With each discovery of exoplanets, our concept of *habitable* expands, incorporating worlds that challenge our understanding of life and its resilience.

Recent discoveries of exoplanets in the so-called *habitable zones* of their stars, thanks to missions such as Kepler and TESS, have amplified these odds. Distant worlds often possess the conditions that could support liquid water, a key ingredient believed to be necessary for life as we know it. Such discoveries increase our expectation of finding life and expand our concept of where life might exist.

From a thermodynamic perspective, we might speculate on the nature of extraterrestrial life. *Pauli's Exclusion Principle* and the *Law of Noncontradiction of Thermodynamics* suggest that life, wherever it is found, must navigate the same fundamental energy constraints.

Recent astrobiology models indicate that life forms based on alternative biochemicals could exist in environments hostile to terrestrial life.

It would have to evolve ways to collect, store, and use energy efficiently, just as life on Earth did. For example, organisms on planets surrounded by thick clouds of interstellar dust could develop photosynthesis based on entirely different wavelengths of light. However, its environment would shape its specific solutions to these problems, demonstrating life's extraordinary flexibility and ingenuity in the universe.

Moreover, James Lovelock's *Gaia theory* suggests that life not only responds passively to its environment but actively modifies it in ways that promote the persistence of life itself. A global homeostatic process could be a universal feature of life wherever it exists. The idea that life is not simply a passenger on the planet but an active agent in shaping its environment offers a fascinating perspective on the possible nature of extraterrestrial life, suggesting that, anywhere in the universe, life would go to work symbiotically with its environment to create conditions favorable to its survival.

Our current technology allows us to search for exoplanets in the habitable zone, where conditions might be suitable for the existence of liquid water. The search for potentially habitable worlds is the first step in understanding the distribution of life in the universe and its favorable conditions. The James Webb Space Telescope, for example, is designed to examine the atmospheres of these planets for signs of life. By analyzing the chemical composition of exoplanetary atmospheres, we hope to detect the fingerprints of life, such as waste gases produced by biological processes.

In the future, more advanced technology might allow us to directly detect signs of life, such as changes in a planet's atmosphere caused by biological activity. We could also send robotic probes to nearby star systems or even use radio or laser signals to try to communicate. Indeed, projects such as *Breakthrough Listen* and *Breakthrough Starshot* strive to overcome our current limitations in detecting signals from extraterrestrial civilizations and designing laser sails that can send small probes to other star systems. Discovering life elsewhere in the universe would have profound implications. It would confirm the cosmic mediocrity hypothesis, which suggests that Earth is not unique in its potential to support life. It would prove that life is a widespread phenomenon, a fundamental part of the thermodynamic strategy of the universe. It would also open new questions, such as whether all life is based on the same principles and how it might continue to defy entropy in different corners of the universe. The possibility of life forms based on completely different principles could inspire new science – an entirely new biology we have not yet conceived of.

9.7 The thermodynamics of alien life

Having considered the possibility of extraterrestrial life, it is intriguing to consider its thermodynamic underpinnings. The vastness of the universe, with its diversity of environments, suggests that life's survival strategies could be as varied as the scenarios in which it might find itself.

Dyson's speculation on radiotrophic trees shows how life on other worlds could use non-solar radiation as an energy source. An idea that expands the concept of photosynthesis to a broader range of electromagnetic radiation suggests that organisms on planets without direct sunlight or in environments exposed to intense cosmic radiation might find alternative ways to capture energy.

Just like life on Earth, alien life would face the challenge of entropy. It must capture, convert, and use energy efficiently, creating structures and mechanisms to keep the relentless tide of disorder at bay. This would involve the evolution of complex self-regulating systems and energy exchange networks optimized for the specific environment in which life develops.

Theories such as the *Fermi Paradox*, which questions why we have yet to encounter alien life forms, suggest that large-scale energy management may be a limiting factor for the visibility of advanced civilizations. This raises the question of how extraterrestrial civilizations can balance the needs of a technologically advanced society with the limitations imposed by entropy and the availability of energy resources.

Although specific characteristics might differ drastically from terrestrial analogs, the same universal laws would apply. This means that, regardless of biochemical or physical particularities, extraterrestrial life forms should operate within the confines of the laws of thermodynamics, adapting and evolving to maximize energy efficiency and minimize entropy production.

The existence of life forms based on entirely different physical principles, as hypothesized by physicists such as Freeman Dyson, who envisions organisms living in the cores of stars or interstellar clouds, should also not be ruled out. If they exist, such life forms could directly exploit nuclear reactions or gas cloud dynamics for their survival, operating under conditions that would be extremely hostile to life as we know it.

The physics and chemistry of such life forms would potentially differ significantly from our own, primarily if they are based on alternative biochemicals, such as those hypothesized to exist on worlds with methane-rich atmospheres, such as Titan, Saturn's moon. In such environments, life structures would rely on solvents other than water, such as ammonia or hydrocarbons, and use energy processes we have yet to conceive of.

What would such structures look like, and how would they function? They could range from simple single-celled organisms that exploit chemical gradients for energy to complex societies that have developed technologies to capture energy from cosmic sources. Such organisms would possess radically different energy transport systems, cell structures adapted to withstand intense radiation or pressure, or even forms of communication based on mechanisms completely different from organic chemistry. Understanding such systems would not only expand our concept of life. However, it could also offer new insights into how life would still manage to defy entropy in different corners of the universe, inspiring new sciences and technologies.

It is hard to say without concrete examples, but we can make assumptions based on thermodynamics. Nobel Laureate Svante Arrhenius hypothesized panspermia, the idea that life can spread through the universe via meteorites, comets, or cosmic dust. This concept suggests that life's survival strategies may be as varied as the environments in which it may find itself, prompting us to consider how different life forms may adapt and thrive in extreme or unusual conditions compared to Earth.

Like terrestrial organisms, alien life forms could evolve in many ways to harness their environment's energy. On a planet with dim, red sunlight, for example, plants might evolve black or dark purple foliage to absorb as much light as possible. In a world with intense volcanic activity, organisms might evolve to harness geothermal energy, much like Earth's extremophiles. Adaptations that illustrate life's flexibility in finding solutions to environmentally imposed energy problems.

Such life forms would also have to manage the waste heat generated by energy use. We might assume they could evolve sophisticated cooling systems or adapt to tolerate high temperatures. The ability to effectively dissipate heat would prove crucial for survival in high-energy environments, suggesting that thermodynamic solutions to heat management are essential for life throughout the universe. Every aspect of their biology, from metabolism to reproduction, should be fine-tuned to operate in a high-entropy environment. An adaptation that not only enables survival but can also drive the development of advanced civilizations that have found innovative ways to harness and manipulate energy to their advantage.

Understanding the thermodynamics of alien life could help us search for extraterrestrial intelligence. If an alien civilization harnesses large amounts of energy – on the scale of a planet, star, or galaxy – it could create detectable heat *signatures*. Futuristic telescopes would be able to detect these infrared emissions, providing evidence of technologically advanced civilizations.

The potential discovery of alien life would give us an extraordinary opportunity to compare how different life forms deal with and manage the universal entropy problem. It would allow us to understand life's universal principles and see how they can be realized in myriad ways, given different environments and evolutionary paths. Although we dream of discovering alien life, we must also reflect on our future, which may be determined by how we manage the struggle with entropy. Our ability to innovate and adapt will allow us to discover and interact with life beyond Earth and ensure the survival and flourishing of human life in the unpredictable and challenging cosmic environment.

9.8 The future of humanity

The thermodynamic principles that have shaped our past and present continue to loom large as we look to the future. From the energy we consume to the technologies we develop, the scope of thermodynamics in shaping our destiny is far-reaching and inevitable. Our ability to understand and apply these principles in innovative and sustainable ways could profoundly affect our future as a species.

Energy remains central to human society, powering our cities, technologies, and industries. As energy consumption increases, so does the entropy we generate, a reality that poses significant challenges and unique opportunities for progress and innovation. How we manage and optimize energy use today will directly impact our future, both environmentally and technologically.

According to the *International* Energy *Agency* (IEA) report, energy efficiency has improved in recent years, but more is needed to offset the increase in global energy consumption. A scenario that highlights the need to accelerate the development of innovative energy technologies and more sustainable consumption practices. The challenge is twofold: there is not only a need to improve the efficiency with which we use energy but also to transform energy sources to reduce our dependence on fossil fuels and minimize the environmental impact of our energy consumption.

Future innovations in energy capture, storage, and utilization will have to confront that reality. Could we develop more efficient solar panels, batteries, or motors? Could we find a way to use waste heat, converting what is currently a nuisance into a valuable resource? Similar questions drive research and development in numerous fields, from physics and engineering to biology and chemistry, with the aim of finding innovative and sustainable solutions.

The thermodynamics of our bodies and minds will also shape our destiny. Nobel laureate Elizabeth Blackburn discovered the importance of telomeres in DNA replication, a critical factor in cellular aging. This discovery suggests new pathways for medical research to combat age-related diseases. Such discoveries can extend human life and offer a deep understanding of how thermodynamic principles operate at the biological level, affecting everything from human health to overall well-being.

From managing our planet's energy resources to finding solutions for human longevity, thermodynamics remains at the heart of our species' most significant challenges and opportunities. As we continue to explore and innovate, the thermodynamic principles that govern the universe will serve as both a guide and a reminder of our limitations and unlimited potential.

Medical advances could enable us to delay the degenerative effects of aging, improve our health, and extend our lifespans. However, no matter how advanced medicine becomes, it will always be constrained by the laws of thermodynamics. Energy can neither be created nor destroyed, and entropy will always increase. Despite our best efforts to combat disease and aging, we must face the limits imposed by the universe's very nature.

Finally, our thermodynamic future is not ours alone. We share this planet with a myriad of other life forms, each of which is struggling with the same fundamental thermodynamic constraints. The interconnectedness among all life forms underscores the importance of considering the ecosystem as a whole when making decisions that affect resource use and entropy management.

Anthropologist and biologist Jared Diamond explored and analyzed how past societies collapsed in part precisely because of their unsustainable use of resources, leaving a warning that resonates today. Lessons learned from the past should guide our future choices, promoting sustainable practices that respect thermodynamic limits and preserve the environment for future generations.

How we interact with our fellow travelers on our thermodynamic journey will significantly impact our collective future. Our choices about resource use, environmental protection, and biodiversity will affect the ecosystem and us. It is essential to take a holistic approach that considers the long-term implications of our actions.

Our thermodynamic choices will broadly define humanity's future. Our understanding of energy and entropy will shape our technological, medical, and ecological decisions. Our ingenuity, innovation, and collaboration ability will be critical in navigating the challenges ahead.

Theoretical physicist Michio Kaku suggests that our long-term survival will depend on our ability to become a *Type II civilization* on the *Kardashev scale*, using energy on a planetary scale. An ambitious goal that will require not only technological advances but also a change in the way we perceive and interact with our environment.

The actions of each of us matter and help determine the future of our species, our planet, and perhaps even life in the universe. By addressing thermodynamic challenges with responsibility and vision, we can build a future in which humanity thrives in harmony with the rest of life on Earth and beyond.

9.9 Immortality

Throughout history, humans have always sought immortality, whether through the elixir of long life of alchemical philosophers or modern biogerontology investigating the mechanisms of aging. Such relentless pursuit reflects humanity's deep desire to transcend the natural limits of life, seeking ways to overcome the cycle of aging and death.

The quest for immortality has been a constant throughout human history, from the myths and legends of ancient civilizations to modern research on aging and life extension. Through the lens of thermodynamics, immortality takes on a new dimension, presenting unique challenges that challenge our understanding of life and the universe.

To understand why, we need to review the principles of conservation of energy and entropy. The basic principles of physics state that while energy in the universe remains constant, entropy, or disorder, tends to increase over time. In a hypothetical scenario in which we achieve biological immortality, such an energy balance would have to be maintained indefinitely, raising profound questions about the sustainability of such an existence.

This could be compared to a perpetual engine, a concept that, despite being disproven by science, captures the collective imagination as a metaphor for the pursuit of endless life. Nevertheless, thermodynamic reality reminds us that every system, including living organisms, is subject to wear and tear and degradation over time due to increasing entropy.

The energy demands of an immortal being in an infinite timeline would become unimaginable. Moreover, the entropy produced would accumulate over time, raising questions about the feasibility of a perpetually sustainable existence in a universe where energy resources are finite and entropy is bound to increase.

The question then becomes: can a finite universe sustain immortal life? This question leads us to reflect on the nature of existence and the limits imposed by the laws of physics. Thermodynamic challenges remain even if we move away from biological immortality and consider digital or artificial forms of existence. Digital conversion of consciousness or evolution to artificial life forms might offer new ways to extend existence beyond biological limits. Still, such life forms would also have to navigate the energetic ecosystem of the universe, facing the same fundamental restrictions of energy and entropy.

As the quest for immortality captures our imagination and pushes the boundaries of science and technology, we remain bound by the immutable laws of the universe. The future of humanity, whether in its biological form or potential digital or artificial expressions, will inevitably be shaped by our ability to understand and operate within fundamental thermodynamic principles.

For example, Ray Kurzweil's work on artificial intelligence suggests that we might one day transfer our consciousness into machines, but even these would need to address the entropy problem. The vision of a future in which the human mind could exist independently of the biological body raises intriguing questions about the nature of being and the continuity of consciousness. Still, it remains constrained by the fundamental physical laws that govern the universe.

Computers and artificial intelligence systems are also subject to the laws of thermodynamics. They need energy to function and produce entropy in the process. A reality that places an inherent limit on the idea of technological immortality and underscores the importance of considering the energy implications of our aspirations toward eternity.

It remains essential to understand that immortality, however tempting, is more than just a technological or medical problem to be solved. It is a deep and fundamental challenge that touches the very principles that govern the universe. The discourse extends beyond physics into fields such as metaphysics and spirituality, questioning whether immortality is desirable or even compatible with human nature.

It is a thermodynamic problem, and as we have seen throughout the book, these problems have challenging solutions. The quest for immortality inevitably confronts the limits imposed by thermodynamics. Our survival relies on a delicate thermodynamic balance: we consume energy to maintain order and produce entropy as a byproduct. This process is essential to life as we know it, and it is an ongoing exploration, as described in the work of pioneers in the field, such as Aubrey de Grey, that pushes the boundaries of biomedicine.

But, if we imagine a hypothetical biological immortality, such a balance would have to be maintained indefinitely, leading to unprecedented energy demands and a continuous accumulation of entropy. The prospect of biological immortality raises practical questions of resources and equity, as access to such technologies could exacerbate social divisions and inequalities.

Such a scenario raises a fundamental question: can a finite universe sustain an immortal entity? The real question is whether we should seek immortality in a universe where change and transformation are the only constant.

The challenge also extends to digital or artificial immortality. With recent breakthroughs in artificial intelligence and deep learning, the possibilities are expanding, but they force us to consider the implications of an identity untethered from biology. Technologies such as computers and artificial intelligence, which could theoretically extend our existence beyond biological limits, are also subject to the laws of thermodynamics.

While technological immortality may seem appealing, it invites us to reflect on the importance of finite human experiences and the intrinsic value of mortality in defining our humanity. Immortality is not simply a technological or medical goal to be achieved but represents a more profound and fundamental challenge.

Facing this challenge forces us to balance the desire to exceed our natural limits with the wisdom of respecting the laws that bind us to the universe. Directly confronting the principles that govern the universe, showing that thermodynamics and immortality are inherently in conflict.

In the words of physicist Richard Feynman, "Nature is not classical...and if you want to make a simulation of nature, you'd better make it with quantum mechanics, and with quantum mechanics, most likely, even an approximation to immortality is not possible."

It is a question that transcends science and touches on philosophical and ethical aspects, forcing us to reflect on the limitations of the universe and the very nature of human existence. Considerations that lead us to recognize the beauty and preciousness of life as we know it, with all its joys and sorrows, victories and losses.

The thermodynamics of immortality presents a profound paradox. On the one hand, it symbolizes the pinnacle of the human challenge against entropy, a potential victory of order and complexity over the inexorable march toward disorder. On the other hand, it represents the dream of alchemists and the desire of countless generations to find the key to transcend the chains of mortality.

On the other hand, it confronts the reality of a finite universe with its immutable laws of energy conservation and increasing entropy. Ultimately, we can find a form of immortality in our continuous desire to understand, improve, and transcend ourselves.

Reconciliation between the human desire for infinity and the limits imposed by the laws of the universe also requires deep philosophical and ethical reflection on our place in the universe.

9.10 The final word

The search for meaning in a seemingly indifferent universe has been the driving force behind many of the most outstanding scientific achievements and philosophical reflections. The constant question of the purpose of life in an environment dominated by entropy drives one to explore beyond the boundaries of human knowledge in search of order and reason in the apparent chaos.

What is the purpose of life in a universe dictated by entropy? Is our existence a mere chance, an aberration in the inexorable march toward thermodynamic equilibrium? Or is there a deeper meaning to it all? These fundamental questions lead us to contemplate our position in the universe and the intrinsic value of our existence.

In the face of the universe's impending heat death, the notion of purpose may seem insignificant, even absurd. Yet, as philosophers such as Thomas Nagel have argued, absurdity does not negate importance; instead, it invites a reconsideration of our approach to existence from a perspective that encourages us to find meaning and value in our lives despite the inexorable advance of entropy.

Yet the very existence of life, in all its complexity and defiance of entropy, suggests an underlying order to the universe – a grand design emerging from the interaction of energy and matter. As we have explored throughout this book, life is a remarkable system of energy dissipation that, through a myriad of complex processes, manages to maintain a delicate balance between order and chaos.

A process observable in the complex *trophic networks* of ecosystems or in the evolution of human societies. It is a balancing act representing not only a biological phenomenon but something universal, reflecting a challenge launched against the so-called *thermal arrest*. In a sense, life can be seen as the universe's strategy to manage energy and entropy. By harnessing energy to create complex structures and processes, life delays the onset of thermodynamic equilibrium, consequently buying time for the universe. It is a survival, resilience, and defiance strategy against the inevitable, demonstrating how each living organism, ecosystem, society, and civilization plays a crucial role in a larger scheme.

This is not a random or meaningless role. Our active participation in the cosmic drama confronts us with the thermodynamics of the universe and challenges us to reflect on mortality, morality, and the search for meaningful life. We are all participants in the Universe's grand design, and exploring and interacting with it may be the key to understanding our purpose.

In the end, whether or not our existence has a purpose may be a question that science can never fully answer, perhaps more of a philosophical or spiritual question.

From the Big Bang to Consciousness

From the Big Bang to Consciousness

Notes and bibliographical insights in Chapter 9:

1. Sean M. Carroll, *From Eternity to Here: The Quest for the Ultimate Theory of Time* (2010). Carroll explores the nature of time through the prism of thermodynamics and quantum mechanics, discussing how the growth of entropy affects our concept of time and the universe's evolution.

2. Brian Greene, *The Fabric of the Cosmos: Space, Time, and the Texture of Reality* (2004). Greene provides an accessible exposition of cosmology and physics theories concerning the nature of space and time, also discussing how entropy plays a crucial role in the universe's physics.

3. Stephen Hawking, *The Theory of Everything: The Origin and* Fate *of the Universe* (2002). In this compendium of his lectures, Hawking discusses the concepts of entropy and various scenarios for the future of the universe, including thermal fate.

4. Roger Penrose, *Cycles of Time: An Extraordinary New View of the* Universe (2010). Penrose proposes a cyclical view of the universe that challenges traditional conceptions of thermodynamic time and introduces concepts that could alter our understanding of the universe's ultimate fate.

5. Freeman Dyson, *Eternal Intelligence in an Expanding Universe* (1979). In this influential essay, Dyson speculates on future intelligence developments in an expanding universe and how it might adapt to or counteract increasing entropy.

6. Frank J. Tipler, *The Physics of Immortality: Modern Cosmology, God and the Resurrection of the Dead* (1994). Tipler explores how the laws of physics could theoretically allow for post-biological forms of life and immortality in a cosmological context.

7. Lawrence M. Krauss, *A Universe from Nothing: Why There Is Something Rather than Nothing* (2012). Krauss discusses the universe's origins from

thermodynamic and cosmological perspectives, examining how entropy and other physical laws influence the formation and fate of the universe.

8. Lisa Randall, *Knocking on Heaven's Door: How Physics and Scientific Thinking Illuminate the Universe and the Modern World* (2011). Randall provides an overview of recent discoveries in physics and cosmology, linking the laws of thermodynamics to the evolution of the universe and emerging technologies.

9. David Deutsch, *The Beginning of Infinity: Explanations that Transform the World* (2011). Deutsch discusses the limits of knowledge and technology, proposing that scientific progress could continue indefinitely despite the limitations imposed by thermodynamics.

10. Max Tegmark, *Life 3.0: Being Human in the Age of Artificial Intelligence* (2017). Tegmark explores how artificial intelligence might evolve to manage resources and entropy in ways that could radically transform our existence and the universe.

From the Big Bang to Consciousness

From the Big Bang to Consciousness

From the Big Bang to Consciousness

PART FOUR

THE INVISIBLE

From the Big Bang to Consciousness

Chapter 10

HIDDEN DIMENSIONS

From the Big Bang to Consciousness

From the Big Bang to Consciousness

> Not only is the Universe
> stranger than we think,
> it is stranger
> than we can think.

WERNER HEISENBERG
Across the Frontiers, 1974

From the Big Bang to Consciousness

10.1 The vibration of wave-particles

We examine the fascinating hypothesis that the Universe is ultimately composed of vibrating waves, a theory corroborated by the astonishing discoveries of quantum physics and the boldest cosmological theories. We imagine the Universe as an order of vibrating elements, where each component, from the most minute photon to the most massive galaxy, plays a specific, indispensable, and related role. Isn't string theory – though not yet scientifically proven – with its mathematically elegant description of vibrating entities in hidden dimensions, a gateway to the unexplored invisible? It launches a bold bridge to ontological questions, proposing quantum interactions not as isolated events but as elements arising from a more hidden dynamic matrix from which existence itself flows.

The structure of spacetime could be intimately intertwined with cosmic vibrations. This perspective incredibly broadens our horizon, going beyond the traditional wave-particle dualism and proposing that mere observation can sculpt reality. For that matter, the collapse of the wave function, the mysterious architect of quantum mechanics, raises profound questions: can the presence of an observer really de-fine and de-limit the existence of the observed?

Meditations that lead us to the threshold of a new answer to the challenge posed by quantum gravity. Heisenberg's uncertainty principle suggests that there are inherent boundaries in our understanding of the Universe, introducing a dose of probability into the very heart of the quantum description of reality. Could reality be far more enigmatic and elusive than our everyday experience suggests? What if becoming is just the tip of a far more complex and multifaceted reality?

In the image of this vibrating universe, it is impossible not to come across the words of Niels Bohr, who stated, "If you are not confused by quantum mechanics, you have not really understood it." The words of mystics and philosophers like Rumi, who saw the universe as a vibrating web of light, no longer seem so distant from modern theories.

10.2 Space-time paradoxes

Albert Einstein's theory of relativity radically transformed our approach to the concept of time, introducing the possibility that it could expand or contract under certain conditions. It was a groundbreaking vision that opened the door to legitimate scientific evaluation of time travel, an idea that until then had traditionally been confined to science fiction novels. In recent times, advanced theories in physics, particularly those related to black holes and event horizons, propose that time paradoxes once considered purely speculative, may have deep roots in the universe's structure. Such paradoxes highlight crucial issues related to the second principle of thermodynamics, which establishes a definite direction for the flow of time, progressing from order toward increasing disorder.

The possibility that *closed time cycles* exist, where future events can influence and modify the past, directly challenges the fundamental principles of causality and logical consistency that govern our physical and philosophical world and raises questions about the existence of free will in a universe where time does not follow a linear, predetermined path. Scientists are exploring the potential of a *theory of everything* that could integrate general relativity and quantum mechanics, offering a more complete explanation of these temporal paradoxes. Such a theory would not only redefine our understanding of time but could also open new frontiers in technology and the exploration of outer space.

The impact of observation on physical reality, a cardinal principle of quantum mechanics, suggests that our interactions with temporal paradoxes may also have direct and unpredictable consequences. Therefore, our epistemological conception of time must be revisited and updated to reflect the new understandings emerging from physical theories. Scientific inquiry is essential in navigating these uncharted waters, paving the way for potential research, technology, and philosophy revolutions.

As we advance on this journey of discovery, we face new ethical responsibilities and incredible possibilities. Manipulating time paradoxes could revolutionize areas such as medicine, allowing us to treat diseases retroactively or the preservation of history, where we may be able to alter historical events for educational or other purposes.

The discussion of the interplay between ancient thought and modern discoveries becomes particularly relevant when we consider St. Augustine's views on time, who admitted his perplexity in being able to define it despite its apparent familiarity. Eastern philosophies on circular causality offer an additional perspective, suggesting that our universe may be far more intricate and interconnected than Western theories have previously assumed. The idea that the future may influence the past, just as the past shapes the future, directly challenges our linear conception of time and opens the door to a more holistic understanding of the universe.

Looking into the future, we can only imagine a world in which temporal paradoxes are understood and actively manipulated, opening up new frontiers in medicine, preserving history, and exploring the universe. This future vision could radically transform our approach to space-time, expanding the range of exploration and enriching our understanding of the interconnectedness of space, time, and matter. The practical applications of such technologies could extend far beyond space travel, affecting areas such as energy, transportation, and communication, exemplifying the importance of pursuing such research for scientific curiosity and the potentially profound impact on our daily lives.

10.3 Holographic Universe

The idea of a *holographic universe* leads us to contemplate one of modern physics's most fascinating and revolutionary theories: that our universe may be a kind of giant hologram where each part reflects the whole. Holographic vision is not just a theoretical exercise; it can radically transform our thoughts about existence. It cannot be ruled out that in the meanderings of this theory, a truth as profound as it is destabilizing may lurk: reality, as we perceive it, is only a shadow of a much more complex and elusive essence.

Holographic reality is a fact that can be hypothesized today from the contradictions detected by scientific observation itself. The deeper one observes matter, the more one discovers that there is no matter, no space, and no locality. If we embrace the idea that the universe is holographic, then what we regard as physical reality could only be a projection of processes occurring on a distant dimensional surface. This means that the concept of locality – the idea that objects are separated in space and time – might be an illusion and that, at a deeper level, everything is connected in ways that are not obvious to our senses. It is an epiphany that forces us to reconsider the very substratum of reality, as such a perspective has staggering implications for our understanding of the universe and our position in it. It is experimentally proven that two initially connected atomic particles once separated even millions of miles apart, continue to behave the same way, just as only the act of observation creates reality. The universe could then be a holographic program designed for our consciousness. Whether we are awake or asleep, we are immersed in a hologram. In such a scenario, the mind is revealed no longer as an isolated entity but as an antenna tuned to infinity. Consciousness could become the sea in which all beings float, bound by invisible currents carrying thoughts, emotions, and knowledge through the space-time web.

Heisenberg's uncertainty principle, rather than limiting our knowledge, opens new doors to understanding the universe as a vibrant entity of infinite possibilities. Observation is not just a passive act but actively participates in creating reality. The mind assumes a crucial role, bridging the infinite potential and the manifest expression of existence.

"Reality is only an illusion, albeit a very persistent one," Einstein teaches us, inviting us to reflect on the ephemeral and constructed nature of what we perceive as reality. An illusion. It is rooted, however, in a foundation at once profound and elusive: *information*. In the holographic universe, information is not merely an aggregate of data but is the fundamental texture from which everything emerges. Consciousness is not just an epiphenomenon of matter but an intrinsic quality of the universe, a vibrant field that permeates and *in-forms* the entity in existence.

The fundamental differences between classical physics and quantum physics can be analyzed in depth by introducing the concept of holographic information. The law of entropy, fundamental to thermodynamics, states that the entropy of an isolated system tends to increase over time, leading to greater disorder or a state of higher probability. In classical physics, the law emphasizes a definite direction of time, from least to highest entropy, reinforcing the idea of a universal, unidirectional arrow of time. Conversely, entropy is closely linked to uncertainty and information in quantum physics. Quantum mechanics, with its properties of superposition and entanglement, suggests that time may not be a universal constant but rather a variable emerging from interactions at the microscopic level. A perspective that radically alters the classical conception of time and entropy, indicating how the understanding of time in quantum physics is intrinsically linked to observation and measurement.

The relationship between entropy and time highlights a fundamental distinction between classical and quantum worldviews. In classical physics, increasing entropy is related to a linear, deterministic progression of time. In quantum physics, however, entropy is understood in terms of information and a measure of uncertainty, introducing additional complexity into the perception of time. Quantum fluctuations and superpositions of states allow, at least temporarily, the entropy increase to evade consideration of the entire system, including observers, raising fundamental questions about the nature of time and the possibility of a less defined arrow of time that is more subject to boundary conditions and initial state.

The holographic principle proposes a revolutionary vision of the universe, where information in a volume of space can be represented on the surface, bounding that volume, suggesting that reality, including the temporal dimension, could emerge from a two-dimensional information substrate. This principle introduces further complexity to the distinction between classical and quantum physics, implying that information, entropy, and space-time are deeply intertwined in a quantum context. The holographic nature of information may offer new insights into the role of time in the universe, suggesting that it may emerge as a property derived from the distribution of information at the fundamental level.

Reflections on entropy, holographic information, and the differences between classical and quantum physics lead to a profound reconsideration of the concept of time. These approaches show how time, in quantum physics, emerges not as an absolute dimension but as a relative aspect that depends on the structure of information and interactions at the microscopic level. The idea that time may have an emergent, non-fundamental nature challenges the traditional conceptions to which we are accustomed and opens the way for new research into the holographic nature of the universe and the very foundations of physical reality.

Time could be an emanation of the universe's fundamental laws governing information and entropy rather than an absolute and independent entity. This understanding would not only resolve some of the paradoxes and open questions in contemporary physics but also guide us to new theories that combine quantum mechanics with the theory of general relativity, offering new avenues for understanding the origin and ultimate nature of time and space.

Exploring the holographic theory is thus an invitation to rethink the nature of the universe and the meaning of every action and experience. If every act, every thought, and every event is woven into the entire universe, then our lives take on enormous meaning and responsibility. Every being and every phenomenon becomes a window to the infinite, an access point to the entire cosmos.

In a holographic universe, the separation between past, present, and future, between here and elsewhere, could only be apparent. Adopting a holographic view would provide a new understanding of consciousness and its relationship to the universe. The violation of the principle of causality and the illusion of time suggest that our linear experience of time may be only a limited perception of a more complex and layered reality. Modern physics and ancient mystical traditions converge on the idea that time, as we know it, is only one aspect of a larger, multidimensional reality in which past, present, and future coexist in an eternal *now*.

"We are not human beings having a spiritual experience. We are spiritual beings having a human experience." Pierre Teilhard de Chardin's phrase profoundly sums up the essence of the holographic universe debate. Let's start with the basics: the external world is just an image produced in our brains. When we see, we see electrical signals produced in the brain. The same is true for all the senses: we assume that the electrical copies in our brain are *reality*.

Quantum information theory offers a tool for understanding how information can exist in a non-localized state, transcending the boundaries of space and time, leading to a view of the universe not as a collection of separate parts but as an indissoluble web of relationships, where each part contains the whole. Perception of reality becomes a decoding of this network of information, a process in which the mind plays a central role. The analogy with the hologram implies that we are illusions created by the surface interference of a coherent wave, a holographic view of existence that challenges us to revise our role in the universe, no longer as external observers but as active participants in a creative process that embraces the entire cosmos.

The challenge posed by the holographic universe is not only scientific but deeply spiritual. It invites us to explore the depth of our consciousness and recognize our fundamental unity with the universe. The idea that the external world is an image produced in our brains, which we see not directly but through electrical signals, prompts us to reconsider the very nature of reality. This is a view that surprisingly aligns with holographic theories, according to which what we perceive as physical reality is, in fact, a projection of processes occurring on a distant two-dimensional surface. Modern science shows us that reality as we experience it is mediated by our senses and brain, leading us to question the veracity of an external world independent of our perception.

The concept of realities as relations, rather than solid objects separated in space and time, is reflected in quantum physics experiments. The notion of a holographic universe as a program created for our consciousness further expands this view into a perspective that implies how *consciousness is not an epiphenomenon of matter but rather a fundamental quality of the universe*, capable of giving shape and meaning to the structure of reality itself. In this context, life appears as a holographic process in which information, or neg-entropy, manifests itself in increasingly complex forms, from inanimate matter to conscious living beings.

The interaction of mind, spirit, and soul in a multidimensional topological context of energy, space, and time suggests a complexity of consciousness that transcends our current scientific and philosophical categories.

Entropy, described so far as a divergent force, could be compared with the opposite concept of *syntropy*, a convergent force reflecting the intrinsic dynamics of the holographic universe. While entropy describes disorder and a tendency toward disintegration, *syntropy could represent a movement toward order, complexity, and, ultimately, consciousness*. This is a fundamental dialectic for understanding the evolutionary nature of reality, where consciousness emerges as an ordering principle within a dynamic universe.

10.4 The law of syntropy

We explore the proposed *law of syntropy*, which suggests increased order and organization in closed systems. This principle could explain phenomena such as cohesion in apparent quantum disorder and self-organization in biological systems.

The term *syntropy* was coined in 1942 by Luigi Fantappiè, one of Italy's most eminent mathematicians, who proposed his unified theory of the physical and biological world. Syntropy would postulate the existence of a dimension symmetrical to entropy, that is, the aspect of universal order. Using principles of relativistic and quantum physics, Fantappiè elaborated how *d'Alembert's equation* could be solved with two opposite solutions: *retarded potentials*, representing waves moving away from a past temporal source, and *anticipated potentials*, waves converging toward a future source. The latter mathematical results, although formally correct, had not previously been considered significant by the scientific community. Fantappiè identified in them a new type of phenomenon that he christened *syntropic*, prevalent in living organisms and constructive phenomena of various kinds.

In the broader context of 20th-century physics and mathematics, these ideas fit into a significant paradigm shift. With the advent of relativity theory and quantum mechanics, methodological certainties were loosening, shifting the axis of physics from a deterministic to a probabilistic and relative science. Heisenberg's uncertainty principle, Gödel's incompleteness theorem, and Einstein's relativism introduced a fundamental element of uncertainty into the scientific domain, opening new doors for understanding the universe's structure.

Syntropy, for its part, introduced a revolutionary element into scientific understanding, suggesting an interplay between the forces of order and disorder that could be essential to understanding phenomena such as growth, evolution, and the functionality of biological organisms. According to Fantappiè, the physical and biological worlds are governed by the tendency of increasing entropy and the possibility of decreasing entropy, i.e., syntropy, which manifests itself in phenomena of greater organizational complexity.

Indeed, Fantappiè's conception of a universe where order emerges from chaos through syntropic processes is not isolated. It reflects and complements the ideas of numerous thinkers and scientists, such as Erwin Schrödinger and his concept of *negentropy* or Teilhard de Chardin's cosmological view regarding psychic energy as an *ordering force*. Contemporary physics, with theories such as that of parallel universes and quantum cosmology, also reflects this duality, recognizing that universal order could result from processes that are both random and determined by initial conditions that favor increasing complexity.

Against this backdrop, syntropy offers an alternative explanation for the mysteries of the biological and cosmic order and stands as a fundamental principle for future scientific and philosophical investigations. Its implications extend beyond the boundaries of traditional science, touching on questions of metaphysics, theology, and philosophy of science and promising to revolutionize our approach to understanding the universe.

10.5 Quantum vacuum structure

The quantum vacuum, traditionally perceived as a vast *reservoir of potential energy*, emerges with such complexity that the idea that it is subtly driven by syntropy, the principle that, as we have said, indicates the intrinsic tendency for the organization of order from chaos, cannot be ruled out. A detailed analysis of how syntropy influences vacuum fluctuations would bring to light previously unexplored layers of energy interaction that form the underlying of our physical reality.

Syntropy, as opposed to entropy, proposes a universe in which even the so-called quantum vacuum is revealed not as a mere absence of matter but as an *active field of unexpressed potentials*. This conception suggests that the quantum vacuum is not a mere vacuum but rather a pulsating environment of activity where energy and matter manifest themselves in ways that can only be indirectly observed through phenomena such as vacuum fluctuations and the *Casimir effect*.

According to this view, vacuum energies, although often regarded as uniformly distributed or statically chaotic, could be ordered following patterns that reflect syntropic principles. Energy patterns could be intrinsically linked to the fundamental structures of the universe, including gravitational and electromagnetic forces. They would offer new interpretations of virtual particle behavior and quantum wave propagation.

Going deeper, if the quantum vacuum is affected by syntropy, we might observe a *hidden order* governing the generation and annihilation of virtual particles and the stability of subatomic structures and cosmological constants. Crucial dynamics to better understand phenomena such as *quantum entanglement* and *superposition*, and even to solve puzzles such as the measurement problem in quantum mechanics.

The implication that the quantum vacuum may function not only as a reservoir of chaotic energy but also as an actively ordered matrix according to syntropic principles raises profound questions about the vacuum's nature and role as a fundamental mediator in the universe. This could radically alter the approach to studying quantum fields, providing new insights into string theory, quantum gravity, and the cosmology of the very early universe.

Exploring the structure of the quantum vacuum in light of syntropy extends our theoretical framework of energy and matter and challenges prevailing conceptions of the origin and evolution of cosmic order. A new approach that would not only entail the possibility of illuminating hidden aspects of theoretical physics but also offer practical implications, such as new technologies based on the control and exploitation of the quantum properties of the vacuum, capable of revolutionizing fields from quantum computation to energy.

10.6 Syntropic force fields

The hypothesis of *syntropy-driven force fields* revolutionizes our approach to the aggregation of matter on a cosmic scale. It raises interesting questions about how these forces can influence the formation of complex structures from simple beginnings. This concept dramatically extends our understanding of phenomena such as crystal growth, galaxy formation, and even molecular self-assembly mechanisms.

Suppose we accept that *syntropic force fields* can operate in nature. In that case, we must consider how these forces can guide processes ranging from atomic arrangements in crystalline solids to the vast architecture of stellar and galactic clusters. Syntropy, characterized by its ability to increase order and reduce entropy locally, could act as a subtle but powerful guide in directing energetic and material interactions that foster complexity and organization.

Syntropic force fields could be responsible for phenomena such as nucleation and crystal growth at the microscopic scale. Typically, these processes are driven by thermodynamic conditions that favor order over disorder. Still, introducing a syntropic principle suggests a more sophisticated and directional regulation of the forces that lead to crystallization. Similarly, in biochemistry and molecular biology, syntropy influences the formation of protein structures and cellular aggregates by operating through force fields that direct the assembly of biomolecular components into functionally optimized configurations.

Moving to larger scales, syntropy could play a crucial role in shaping the aggregation dynamics of matter in star and galaxy formation processes. Structures such as galaxies, which exhibit a high degree of order and structure despite the dominant influence of entropy at the cosmic level, could result from syntropic fields working in tandem with gravity to influence the distribution of matter and dark energy.

Additionally, these syntropic force fields offer a new lens to examine and unify seemingly disconnected physical phenomena, such as nature's fundamental forces and cohesion mechanisms in advanced composite materials. Their study could lead to discovering previously incomprehensible fundamental interactions, thus offering new ways to exploit these natural phenomena technologically.

Introducing syntropic force fields into scientific discourse would enrich our understanding of natural and artificial processes and inspire new technologies based on more effective and precise manipulation of fundamental forces. The implication that order might emerge spontaneously in response to these implied forces suggests an entirely new paradigm for materials engineering, theoretical physics, and cosmology.

10.7 Collective consciousness

As part of the study of the interactions between syntropy and consciousness, it is possible to explore the hypothesis that syntropy acts as a catalyst in the emergence of collective consciousnesses or global neural networks with significant impacts on the formation of collective thought and intelligence at higher levels of organization. This investigation is part of a broader context linking *physics to deep psychology*, primarily through the concept of the *collective unconscious* formulated by Carl Gustav Jung.

Syntropy, described as a force that promotes order and cohesion against entropic chaos, could be fundamental in structuring not only matter but also information and human interactions on a large scale. In the social and cultural context, syntropy would make the formation of collective consciousness easier, influencing how ideas and behaviors propagate and stabilize within communities.

Jung's concept of the collective unconscious is based on the existence of archetypes, innate mental structures shared by all human beings. In this context, syntropy could be seen as a force that drives the aggregation and manifestation of these archetypal structures in ways that emerge in the collective consciousness. It could influence, for example, the rapid spread of cultural and technological innovations that seem to arise simultaneously in different places in the world without an apparent direct connection between them.

The interplay between syntropy and the collective unconscious could be crucial to understanding phenomena such as major social and cultural change waves. In epochs of intense transformation, such as those that have characterized the transition to the digital age, underlying syntropic dynamics would be present that facilitate a collective restructuring of cultural priorities and intellectual paradigms.

Moreover, the introduction of digital technology and global communication networks could be seen as a physical extension of these syntropic force fields, with the Internet serving as a tool for the physical manifestation of the collective unconscious. Networks not only transmit information but can also modulate and direct the formation of collective consciousness through mechanisms that reflect the syntropic principles of organization and connectedness.

At the neuroscientific level, the idea that syntropy can influence the formation and functionality of neural networks suggests potential applications in fields such as artificial intelligence and robotics. Understanding how syntropy can optimize neural networks would lead to significant developments in creating AI systems capable of emulating the distinct human ability to adapt and carry out collective learning.

10.8 Toward a holistic understanding of the Universe

In this chapter, we have explored various deeply interconnected aspects of the universe, from its fundamental structure to the broader implications of collective consciousness, via the revolutionary paradigms of modern physics. We have seen how syntropy, a hypothetical force that promotes order and cohesion from chaos, can play a crucial role in phenomena beyond the traditional understanding of classical physics, suggesting a far more interconnected and dynamic reality than previously imagined.

We began with the hypothesis that the universe can be understood as an order of vibrating elements, a view supported by string theory and recent discoveries in quantum physics. A concept that reinforces the idea of a universal substrate in which every component, from subatomic particles to galaxies, is intrinsically connected in a dynamic exchange of energy and information.

Einstein's theory of relativity and subsequent developments in black hole physics introduced the notion that time can be flexible, paving the way for theories that allow for time paradoxes and time travel once relegated to science fiction. Concepts that challenge our understanding of causality and raise profound questions about free will and the nature of time itself.

The proposal that our universe may be a holographic projection of processes on a distant two-dimensional surface upsets our conceptions of space and locality. A perspective that suggests how reality and information are non-localized and that each part of the universe is deeply connected to every other, offering a radical vision of cosmic interconnectedness.

Syntropy is an organizing principle that offers a framework for understanding how order can spontaneously emerge from chaos. This concept can be applied to the formation of crystal structures, the dynamics of galaxies, and the behavior of neural networks.

Thus, we considered how syntropy could facilitate the emergence of collective consciousness through global neural networks, influencing the formation of collective thought and intelligence.

In conclusion, this chapter presents a view of the universe not as a series of isolated and independent events but as an intricate web of interdependent dynamics, where syntropy serves as a catalyst for an emerging order.

From the Big Bang to Consciousness

Notes and Bibliographical Insights to Chapter 10:

1. Sean M. Carroll, *From Eternity to Here: The Quest for the Ultimate Theory of Time* (2010). Carroll deepens the understanding of time through thermodynamics and quantum mechanics, discussing the role of entropy in the perception and temporal evolution of the universe.

2. Brian Greene, *The Fabric of the Cosmos: Space, Time, and the Texture of Reality* (2004). In this accessible book, Greene explores the theories that describe the structure of space and time, including the concepts of entropy and the holographic universe.

3. Stephen Hawking, *The Theory of Everything: The Origin and Fate of the Universe* (2002). Hawking discusses entropy and the universe's future, proposing theories about its holographic origin.

4. Roger Penrose, *Cycles of Time: An Extraordinary New View of the Universe* (2010). Penrose explores a cyclical approach to the universe that could reformulate the understanding of time and introduces theories that align with the vision of a holographic universe.

5. Michael Talbot, *The Holographic Universe* (1991). Talbot presents the theory that the universe is structured as a hologram, where each part contains the whole, leading this view back to research in quantum physics and neurophysiology.

6. Luigi Fantappiè, *On a New Theory of "Ultimate Relativity"* (1942). This is the original article by Fantappiè, which introduces the concept of syntropy, proposes a complementary view of entropy and focuses on order and cohesion emerging in systems.

7. Luigi Fantappiè, Unitary *Theory of the Physical and Biological* World ((1942). Fundamental text introducing the concept of syntropy and describing Fantappiè's Unitary Theory of the Physical and Biological World.

8. Ervin Laszlo, *Science and the Akashic Field: An Integral Theory of Everything* (2004). Laszlo provides an overview of theories that view the universe as a holographic information field, tying syntropy to cosmic memory or the Akashic field.

9. Lawrence M. Krauss, *A Universe from Nothing: Why There Is Something Rather than Nothing* (2012). Krauss examines the origins and destiny of the universe using thermodynamics and cosmological theories, even touching on the possibility of a holographic universe.

10. David Bohm, *Wholeness and the Implicate Order* (1980). In this work, Bohm proposes a holographic universe model in which reality is seen as a whole.

11. Werner Heisenberg, *Physics and Philosophy: The Revolution in Modern Science* (1958). Heisenberg discusses the implications of the uncertainty principle, which revolutionized modern physics by introducing fundamental uncertainty into the scientific realm.

12. Kurt Gödel, *Über formal unentscheidbare Sätze der Principia Mathematica und verwandter Systeme I* (1931. Gödel presented his incompleteness theorem here, showing the limits of mathematical logic and contributing to a significant paradigm shift.

13. Paul Davies, *The Accidental Universe* (1982)Davies investigates the laws of the universe that foster organizational complexity and order, offering insights into the possible holistic nature of the universe.

14. Erwin Schrödinger, *What is Life? The Physical Aspect of the Living Cell* (1944). Schrödinger introduces the concept of negentropy and explores its implications for biological systems, suggesting that life thrives on negative entropy.

15. Pierre Teilhard de Chardin, *The Human Phenomenon* (1955). Teilhard de Chardin discusses how psychic energy acts as an ordering force in the cosmos, integrating the spiritual and material aspects of the universe.

16. Hugh Everett, *The Many-World Interpretation of Quantum Mechanics* (1957) is Everett's seminal work on many-world interpretations. It postulates parallel universes as a framework for understanding quantum phenomena.

17. Stephen Hawking, George F.R. Ellis, *The Large-Scale Structure of Space-Time* (1973). This book provides a comprehensive account of the theoretical basis

needed to understand the dynamics of large-scale space-time structures within the framework of general relativity.

18. Leonard Susskind, *The Cosmic Landscape: String Theory and the Illusion of Intelligent Design* (2005), discusses how string theory and the concept of a cosmic landscape can provide a deeper understanding of the universe's fine-tuning.

19. Roger Penrose, *The Road to Reality: A Complete Guide to the Laws of the Universe* (2004). Penrose explores the relationship between the fundamental laws of physics, including thermodynamics, and the universe's structure.

20. Carlo Rovelli, *Reality Is Not What It Appears To Us: The Journey to Quantum Gravity* (2014). Rovelli offers insights into how quantum gravity theories could bridge the principles of general relativity and quantum mechanics, highlighting the role of informational and entropic processes in the fabric of spacetime.

21. Max Tegmark, *Our Mathematical Universe: My Search for the Ultimate Nature of Reality* (2014). Tegmark examines the hypothesis that physical reality is a mathematical structure that inherently includes a synthesis of entropy and its opposing forces.

22. Carl Gustav Jung, *The Ego and the Unconscious* (1928). Jung explores the concept of the collective unconscious and introduces archetypes as central elements of the human psyche, fundamental to understanding the formation of collective consciousness.

23. Ervin Laszlo, *Cosmos: An Integral Worldview* (2007). Laszlo discusses how modern physics, including the concepts of quantum field and universal interconnectedness, can be aligned with psychology to explore collective consciousness.

24. Fritjof Capra, *The Web of Life* (1996). Capra analyzes how biological systems exhibit a complex interdependence that can be compared to the processes of collective consciousness formation.

25. Rupert Sheldrake, *The Extended Mind* (2013). Sheldrake introduces the concept of morphogenetic fields, suggesting that non-material mechanisms influence the formation of collective consciousnesses.

26. Brian Greene, *The Elegant Universe* (1999). Greene explains how string theory can provide a unified view of the fundamental universe's fundamental forces and the holistic model that interconnects all universe components.

27. Stephen Hawking, Leonard Mlodinow, *The Big Picture* (2010). Hawking and Mlodinow discuss how modern physics, including M-theory, suggests a holistic view of the universe where chaos and order are interconnected.

From the Big Bang to Consciousness

CONCLUSION

Life is not simply a passive observer of the universe's narrative but a central figure heroically battling the invincible force of entropy. In the cosmic epic, every living being becomes a co-author, ceaselessly shaping the narrative with new discoveries and inventions. The intimate connection between conscious life and cosmic infinity leads to a journey of enlightenment that spans the various dimensions of existence, from the intricate microcosm of DNA to the immense macrocosm of galaxies.

The principles of thermodynamics once thought to be implacable antagonists to life's existence, have become vital agents that enable and shape the nature of life itself. This is a narrative reversal that transforms supposed enemies into indispensable allies. Understanding this concept illuminates the elegant paradox of our existence: life is not a fluke in an indifferent universe but a natural result of the laws of the universe, which are, in part, yet to be understood.

In the fabric where the fundamental laws of the universe and the evolutionary germination of being intertwine, we come across an exemplary manifestation of resilience. Such resilience transcends the mere biological dimension, rising to an ontological principle that imbues the essence of human existence. A motif incessantly investigated through art and literature that exalts the potential of being to flourish in the mute adversity of fate. Life, in its limitless manifestations, thus emerges as the mode by which the universe counters the inexorable advance toward thermodynamic equilibrium, resisting and even triumphing in the face of the impositions of the second law of thermodynamics. In every living entity, from the cellular microcosm to the sophisticated human encephalon, the cunning of exploiting said principle to establish an order capable of sustaining self-knowledge and self-consciousness predominates.

We have scrutinized DNA, the language of life, understanding its role as a blueprint for life and a thermodynamic code that mediates energy flow. This understanding amplifies our admiration for the *coded language of* nature, as geneticist George Gamow called it. The thread extends to our exploration of memory and the human mind, uncovering the connections between our thoughts, experiences, and the fundamental laws that govern the cosmos.

By exploring the thermodynamics of societies, cultures, and technology, we have seen how human constructs mirror the patterns of the larger universe. This echoes the principle that every system, whether an atom or a society, follows the same fundamental laws of organization and chaos.

As we stand on the brink of a largely unknown future, the insights gleaned from our exploration illuminate the way forward. In its myriad forms, life appears to be the universe's way of organizing itself against the march toward equilibrium, persisting and thriving in the face of the second law of thermodynamics.

Each life process becomes part of a larger cosmic narrative, a testament to life's resilience and inventiveness. Philosophically, this vision echoes the thinking of Teilhard de Chardin, who saw evolution as a process directed toward a point of maximum complexity and consciousness, the so-called *Omega Point*.

Already in ancient Greece, philosophers such as Empedocles reflected on the tension between order and chaos, an idea that is reflected in modern theories of thermodynamic non-equilibrium. Speculative theories such as those of information process physics suggest that life may be fundamentally a computational phenomenon.

But there is something that transcends the very concept of computation. Even in its most apparent chaos, there is a certain rhythm, an aesthetic that arises from the need for survival and the drive toward complexity. The splendor of a blooming flower, the melody of a singing bird, the vibration of a bustling city, and the glow of a distant star all testify to the beauty of cosmic choreography. Such a spectacle evokes the *sensus divinitatis* spoken of by philosopher John Calvin, indicating a profound resonance between nature and the human perception of the divine.

In nature, entropy governs the cycle of life and death, the constant ebb and flow of the seasons, and the sculpting of landscapes through erosion and weathering. A phenomenon exalted in the aesthetic concept of *wabi-sabi* in Japanese culture, which finds beauty in imperfection and impermanence. When we see the breathtaking panorama of a starry night, we witness a cosmic theater in which entropy is the director, guiding the life cycle of stars from their fiery birth to their explosive end.

Art, in many ways, echoes this. Artists are often inspired by the beauty of decay, time passage, and the struggle between order and disorder. Just think of the work of artists such as Salvador Dalí, whose surrealist interpretation of time and reality challenges our perception of order. A piece of clay can be molded into a beautiful sculpture, a blank canvas transformed into a visual feast of colors and shapes, and a silent air vibrates with the harmonious melody of music. The influence of entropy in art can be compared to the distillation process in literature, where meaning is created through the reduction and concentration of ideas. The artist's challenge is to create order, meaning, and beauty in the face of the natural tendency toward randomness and disorder.

Delving into life's grand design, one cannot help but marvel at the profound beauty and harmony underlying everything. Such a feeling of awe has been expressed through the centuries by philosophers and mystics like Rumi, who saw the universe as a manifestation of divine love, and by scientists like Einstein, who perceived in recognition of the order of the universe an experience of the sacred. Awareness, in turn, opens the door to exploring the spiritual dimensions of thermodynamics. It is not a matter of adhering to a particular religious doctrine but instead of recognizing the profound sense of interconnectedness and awe that the thermodynamic view of life generates.

The spiritual dimensions of thermodynamics flow from the recognition that we are all part of a complex order of energy and matter, constantly moving between order and chaos, complexity and simplicity, life and death. A concept that finds parallels in the Hindu idea of *Brahman*, the omnipresent cosmic spirit, and in the scientific notion of the *quantum field* that permeates the entire universe. Similar reflections are found in Taoism, where harmony with the Tao implies an existence in balance with the natural cycle of things. Such a perspective encourages us to embrace a holistic view of life, recognizing the sacred in every life form and every grain of sand, as each element plays its role in the universe's challenge against entropy.

Some modern speculative theories, such as those involving quantum physics, suggest that there may be an as-yet unexplored connection between consciousness and matter. These theories offer new perspectives on life as a biological and physical phenomenon. Many open questions and doubts, such as the ultimate fate of the expanding universe and entropy's role in the long term, continue to stimulate imagination and scientific research.

With the advent of quantum physics, physics is no longer just the study of motion and energy but becomes an exploration of the *very foundation of* existence. The evolution from mechanistic physics to quantum physics marks an epistemological transformation: from a world of tangible particles to a realm where the particle exists only when observed, escaping being when unperceived, transforming into vibration, a wave of probability.

To study quantum physics, then, is to immerse oneself in a domain essentially made up of shadows and probabilities, where events occur not by iron causality but by statistical possibility. A perspective that has radically eliminated the classical concepts of space, time, and locality, leading us to confront *non-locality* – a principle that challenges our common sense of separation and distance.

This epistemic explosion requires us to reconcile the opposing dualities of the pre-Socratic philosophers Parmenides and Heraclitus. Parmenides saw existence as a *single*, unchanging *block*, denying change and movement as illusions. On the contrary,

Heraclitus proposed that everything is in a constant *state of flux*, where "one can never bathe twice in the same river." These opposing philosophical reflections can be reconciled through the nature of quantum reality, where what we observe depends on our act of observation.

The Eastern mystical view, with its concepts of Maya, cosmic illusion, and Sunyata, vacuity, further reinforces the idea of an intrinsic reality that is atemporal and aspatial. In this conception, the distinction between the manifest and the non-manifest is a product of human perception, not an inherent condition of the universe.

The insights we gain about the principles of the universe and how the design of life aligns with them allow us to look forward with a new sense of purpose. It is a call to reflect on our position in the universe, a testament to its ingenuity, and a tribute to life's resilience.

It is an invitation to understand, experience, and evolve – *a way for the universe to recognize itself.*

From the Big Bang to Consciousness

From the Big Bang to Consciousness

Index of main authors

Aiello, Leslie: 181
Arbesman, Sam: 238
Aristotle: 17, 63, 120, 149
Arrhenius, Svante: 280
Augustine, St.: 149, 303

Barrett, Lisa Feldman: 182
Benveniste, Jacques: 155
Bergson, Henri: 158-160
Blackburn, Elizabeth: 282
Bohr, Niels: 301
Bostrom, Nick: 26

Calvin, John: 323
Cassirer, Ernst: 209
Chomsky, Noam: 210
Crick, Francis: 91

Dalí, Salvador: 323
Darwin, Charles: 63, 74, 119, 176, 269
De Grey, Aubrey: 286
Dennett, Daniel: 137, 157
Dewey, John: 175
Diamond, Jared: 215, 283
Drake, Frank: 277
Durkheim, Émile: 206
Dyson, Freeman: 255, 276, 279-280

Einstein, Albert: 252, 302, 304, 308, 314
Empedocles: 322
England, Jeremy: 223

Fantappiè, Luigi: 308-309
Feynman, Richard: 287
Foucault, Michel: 213-214
Friedman, Thomas L.: 219
Frijda, Nico: 182
Friml, Jiri: 155
Friston, Karl: 180, 183-184

Gamow, George: 322
Gigerenzer, Gerd: 184
Gutenberg, Johannes: 120

Hameroff, Stuart: 179, 187
Hanson, Robin: 274
Hardin, Garrett: 235
Hawking, Stephen: 265-269
Hegel, Georg Wilhelm Friedrich: 19
Heidegger, Martin: 233
Heisenberg, Werner Karl: 301, 304, 308
Heraclitus: 324

Hodgkin, Alan: 176
Homer-Dixon, Thomas: 216
Humphrey, Nicholas: 181

Jaynes, Edwin Thompson: 181
Jevons, William Stanley: 257
Jonas, Hans: 140, 249
Joyce, Gerald: 66
Jung, Carl Gustav: 312

Kahneman, Daniel: 183
Kaku, Michio: 284
Kandel, Eric: 177
Kant, Immanuel: 31
Kauffman, Stuart: 237
Kolbert, Elizabeth: 223
Kurzweil, Ray: 187, 285

Latouche, Serge: 221
Laughlin, Simon: 178
Lévi-Strauss, Claude: 208
Llinás, Rodolfo: 181
Locke, John: 150
Lovelock, James: 278

Marx, Karl: 206
Metzinger, Thomas: 162
Michelangelo: 120
Miller, Stanley: 63, 65
Morin, Edgar: 252

Nagel, Thomas: 288
Nash, John: 183
Newton, Isaac: 31

Nowotny, Helga: 238
Nussbaum, Martha: 182

Ostrom, Elinor: 212

Parmenides: 324
Penrose, Roger: 164, 179, 187
Piaget, Jean: 185
Plato: 13-14, 17, 149
Popper, Karl: 236
Prigogine, Ilya: 37, 44, 68, 71, 177, 205

Ramón y Cajal, Santiago: 175
Rifkin, Jeremy: 208
Rumi, Jalal al-Din: 301, 323

Sapolsky, Robert: 182
Schrödinger, Erwin: 33, 175, 256, 309
Schumpeter, Joseph: 217
Selye, Hans: 185
Shakespeare, William: 120
Shannon, Claude: 210, 242
Spencer, Herbert: 119
Spengler, Oswald: 216
Susskind, Leonard: 241
Szostak, Jack: 66

Tainter, Joseph: 215
Taleb, Nassim Nicholas: 68, 216
Teilhard de Chardin, Pierre: 47, 139, 306, 309, 322
Tononi, Giulio: 179

Urey, Harold: 63, 65

Varela, Francisco: 138

Walker, John E.: 67
Watson, James: 91
Weart, Spencer: 222
Wheeler, John: 164
White, Leslie: 205
Whitehead, Alfred North: 43
Whorf, Benjamin Lee: 211
Wiśniewska, Justyna: 156

Yunus, Muhammad: 218

Zalasiewicz, Jan: 140

From the Big Bang to Consciousness

Index of main concepts

A

Abiogenesis: 64-66, 69-71
Adjacent possible: 237-238
Anomie: 218
Anthropic principle: 34, 65
Anthropocene: 137, 222-224, 236, 248
Antifragility: 68, 216-217
Arrow of time: 41, 64, 120, 122, 188, 255, 305
Autopoiesis: 138

B

Beneficial mutations: 79, 125
Biodiversity: 78, 125, 129-131, 138-139, 222-223
Biogeochemical cycles: 99
Bipedalism: 136
Brain entropy: 159, 161, 179-180

C

Catastrophic complexity: 215
Chaos theory: 64, 68, 188, 209
Chemical gradients: 73, 280
Circular economy: 224, 247, 257
Co-evolution: 130-131, 139
Cognitive sophistication: 133
Collective unconscious: 312-313
Critical point: 205, 224, 271
Cultural thermodynamics: 210

D

d'Alembert's equation: 308
Dark energy: 69, 187, 311
Degrowth: 221, 249, 257
Determinism: 135, 188
Dissipative structures: 37, 44-47, 68-73, 76, 79, 205
Dynamic reorganization: 159

E

Ecological footprint: 234, 246
Emergent order: 71, 98
Encephalization: 180
Encoding of memories: 151
End of time: 265
Endosymbiosis: 70, 74
Energy economy: 68, 94, 160, 207
Energy value theory: 205
Entanglement: 305, 310
Epigenetics: 99-101
Evolutionary epistemology: 236
Extremophiles: 63, 65, 270, 281

F

Fermi paradox: 279
First law of thermodynamics, the: 35, 205
Free will: 135, 188, 302, 314

G

Gaia hypothesis: 34, 278
Game theory: 183, 211, 218
Gene flow: 118
Genetic drift: 118
Gibbs free energy: 36, 93
Gini index: 207
Gossen's economic law: 159
Gravitational wave: 69
Guided dissipation: 223

H

Helmholtz free energy: 36
Holographic principle: 71, 305
Hubble's law: 268

I

Informational entropy: 92
Ionic gradients: 151-152
Internet of Things (IoT): 245, 251
Isolated systems: 36, 40, 94

K

Kardashev law: 256
Kardashev scale: 284

Koomey's law: 186, 249

L
Landauer limit: 250, 256
Linguistic entropy: 211

M
Memory consolidation: 150
Metabolism: 46, 63, 70, 75, 78-79, 93, 101, 177, 270, 281
Moore's law: 243, 249
Moravec's paradox: 186, 220
Mutation: 77- 81, 91, 101-103, 117, 124-127, 270

N
Negative entropy (Negentropy): 34, 256, 309
Neurogenesis: 176
Neuroplasticity: 134, 156, 175

O
Omega Point: 47, 322
Orchestrate Objective Reduction: 187

P
Panspermia: 75, 138, 271, 280
Phenotype: 100, 125
Photosynthesis: 46, 48-50, 120, 123, 270, 278-279
Presentism: 157
Primordial soup: 63-73
Principle of conservation: 35

Q
Quantum mechanics: 32, 250, 287, 301-302, 305-306, 308, 310

R
Recombination: 97-98, 103, 117
Redundancy: 96, 242
Reversible computation: 188

S
Schrödinger's paradox: 33
Second law of thermodynamics, The: 17, 34, 35-37, 42, 44, 49, 64, 67, 92, 94, 102, 127, 163, 177, 188, 206, 217, 219, 244, 265, 268, 271, 322
Self-organization: 50, 66, 69, 72-74, 138, 308
Social homeostasis: 206
Systems science: 205
Synaptic plasticity: 150-154
Syntropy: 308-314
Syntropic force fields: 311-312

T
Theory of Everything: 302
Third law of thermodynamics, The: 36
Third Industrial Revolution: 208

U
Unified field theory: 68

W
Wasted energy: 177, 242

From the Big Bang to Consciousness

From the Big Bang to Consciousness

From the Big Bang to Consciousness

From the Big Bang to Consciousness

Although the author has made every effort to ensure that the information contained in this book is correct, and although the publication is designed to provide an accurate exposition of the subject matter covered, the author assumes no responsibility for any errors, inaccuracies, omissions, or any other inconsistencies herein and hereby disclaims any liability to any party for any loss, damage of any kind caused by errors or omissions, whether such errors or omissions arise from negligence, accident, or any other cause.

www.ingramcontent.com/pod-product-compliance
Lightning Source LLC
Chambersburg PA
CBHW052239220526
45471CB00001B/111